Biotechnology and the Politics of Plants

Biotechnology and the Politics of Plants explores the mysterious phenomenon of 'apomixis', the ability of certain plants to 'self-clone', and its potential as a revolutionary tool for agriculture and enhancing food security, that may soon be a reality. Through historical anthropological and ethnographic study, Matt Hodges traces the development of the CIMMYT Apomixis Project, a prominent frontier research programme, and its reinvention as a leading public-private partnership (PPP). He analyzes the fast-moving historical transition from public sector, mixed plant breeding approaches grounded in genetics, to a contemporary era of agricultural biotechnology and genomics where PPPs are a leading format. In doing so, he explores how social contexts of research shape how knowledge is produced, as well as what remains 'unknown', and constrain the development of an 'Apomixis Technology'. The book presents an inventive approach informed by the anthropology of time, science and technology studies, and dialogues with the work of Gilles Deleuze, Paul Rabinow, Hannah Arendt, Andrew Pickering and Eduardo Viveiros de Castro. Hodges outlines novel ways of integrating notions of history and becoming, and considers how apomixis offers up an alternative image of thought to theoretical concepts such as the well-known 'rhizome'. *Biotechnology and the Politics of Plants* makes a valuable contribution to both the growing social scientific literature on genomics and biotechnology, and recent anthropological debates on time and history.

Matt Hodges is a social and historical anthropologist based at the School of Anthropology and Conservation at the University of Kent, UK. He works on the anthropology of science and technology, and themes of history, time and the experience of cultural transformation and rupture in rural Europe. This focus extends to the technologies and infrastructures that drive such upheavals, including agricultural biotechnology. Recent work on French radical historiography appeared in *Current Anthropology* 60(3).

Routledge Focus on Anthropology

The Primate Zoonoses
Culture Change and Emerging Diseases
Loretta A. Cormier and Pauline E. Jolly

Aerial Imagination in Cuba
Stories from Above the Rooftops
Alexandrine Boudreault-Fournier

Biotechnology and the Politics of Plants
Disciplining Time
Matt Hodges

For more information about this series, please visit:
www.routledge.com/anthropology/series/RFA

Biotechnology and the Politics of Plants
Disciplining Time

Matt Hodges

Routledge
Taylor & Francis Group

LONDON AND NEW YORK

First published 2021
by Routledge
2 Park Square, Milton Park, Abingdon, Oxon OX14 4RN

and by Routledge
605 Third Avenue, New York, NY 10158

Routledge is an imprint of the Taylor & Francis Group, an informa business

British Library Cataloguing-in-Publication Data
A catalogue record for this book is available from the British Library

Library of Congress Cataloging-in-Publication Data
A catalog record has been requested for this book

ISBN: 978-1-138-31452-8 (hbk)
ISBN: 978-0-367-20146-3 (pbk)
ISBN: 978-1-003-18309-9 (ebk)

Typeset in Times New Roman
by Newgen Publishing UK

For Lorea—with love.

Contents

Figures

Acknowledgements

A book of this kind emerges from a nexus of relations—which I now gratefully acknowledge. My sincere thanks to the CGIAR Secretariat for their invitation to participate in the CGIAR Alliance Deputy Executive and Private Sector Committee Workshop on Public-Private Partnerships, at which I presented key arguments and conclusions of this research. An outline of the principal findings of part 3 was first published on the CGIAR website as electronic output from the workshop, as were the visuals. I would also like to thank CIMMYT for the kind invitation to deliver an anthropological study of the projects in this book for that event, and vital support in providing access to ethnographic information. CIMMYT also kindly facilitated the use of its library in Mexico, and this openness reflects that institution's commitment to provision of international public goods. I am also grateful to Steve Hughes and Andrew Pickering from my time at the University of Exeter, whose support for this work was greatly appreciated. Steve Hughes, as a co-signatory of the Bellagio Apomixis Declaration, was a firm believer in the potential of apomixis for agriculture. Andrew Pickering's guidance was indispensable, as was the warm welcome he offered into the field of Science and Technology Studies. I was also fortunate to interview selected participants and stakeholders in these projects and, where relevant, from the wider field of apomixis research. Invariably, the welcome I received was open and friendly, even if there were sometimes limits to the information that could be shared. I do not name these individuals here, but I thank them for sharing their extensive knowledge, opinions and insights into the subject matter of this book, and for their encouragement.

Turning elsewhere, some of this research was presented as part of the Economic and Social Research Council (ESRC) Seminar Series, 'Conflicts in Time: Rethinking "Contemporary" Globalisation', convened by Laura Bear and Stephan Feuchtwang. I would like to thank Laura and Stephan for their kind invitation, and for their encouragement and constructive feedback on the work I presented over the course of that series of meetings. My warm thanks to Glenn Bowman and Elizabeth Cowie for their friendship, advice and for their hospitality, which was always invaluable. Charles Stewart is a source of ongoing support and inspiration for my work on time and history, and my thanks and appreciation to him, and to Stephan Palmié and Eric Hirsch, who have encouraged my work on the anthropology of history. Charlie and Patricia Bonallack offered hospitality and good company at their home in Montpellier during my research in France. At the University of Kent, I am fortunate to have colleagues—Anna Waldstein, Dimitrios Theodossopoulos, Ozlem Biner, and Raj Puri—whose research is an inspiration. Conversations with my PhD students Jade Richards, Oscar Krüger, and Chris de Coulon Berthoud were illuminating. Tracy Kivell helped make time for me to complete this book across a dark summer, which I greatly appreciate, and my warm thanks also to Alexandra Leduc-Pagel for her enthusiasm and support for this project. And I am grateful to Judith Bovensiepen, for her many kindnesses and encouragement during the writing of this book, and for sharing her anthropological acumen when I needed it most. Of course, no one named above is responsible for the content of the narrative that follows. At home in Canterbury, I would like to thank Susana Gozalo Martínez for her warm support, and many other things without which this project could not have been completed. Finally, Niko Hodges and Lorea Hodges deserve a special thank you, as they waited patiently for this book to be completed over a long lockdown, and still found plenty to smile about in the sunshine.

Note on the text

This book is a work of cultural anthropology. Pseudonyms replace the names of selected organizations, and some acronyms which are used to refer to them, in line with anthropological best practice and in order to protect their identity. For ethical reasons to respect the privacy of real people, some professional, personal and geographical details are also changed, as are some anthropologically inconsequential details of the narrative and of the main projects, including names. In keeping with the aims of anthropological discourse, this account operates at the level of ethnographic and cultural redescription. This means that it draws on real events as sources of empirical and ethnographic information, but transposes them into anthropological narrative, which operates for academic and disciplinary ends. It is not a journalistic account and no one should mistake the representations of individuals that follow for objective representations of real people. Confidentiality agreements restricted research access to certain information, as is to be expected where commercial interests are at stake. Where the narrative concerns major state or non-governmental institutions, or multinational corporations, I have usually left real names in place as points of reference. Key research was undertaken as part of the work of the ESRC Genomics Network, and funded by ESRC grant RES-145-28-0001. I would like to thank the ESRC and the Department of Sociology and Philosophy at the University of Exeter where I was originally based. Later research and writing was supported by the University of Kent.

1   The politics of emergence

The quest for an Apomixis Technology

For more than 60 years plant breeders and scientists have sought to create the holy grail of modern plant breeding—a crop plant that can clone itself. A mysterious process known as *apomixis* exists among plants as familiar as the dandelion, whose hieroglyphic aura of concealed meaning science aims to penetrate (van Dijk et al. 2009). Such 'wild' plants can reproduce asexually and release seeds containing a copy of the maternal plant's DNA, which grow into a 'clone' of the original plant. Introduce this into maize or wheat, scientists propose, and you will forge a crop plant that can self-clone—and revolutionize the way in which humans grow their food (Hojsgaard 2020; Spillane et al. 2004). From the Soviet Union to Mexico, India and Australia to the United States, this quest has stirred the imaginations of scientists, plant breeders, governments, corporations and wealthy funders alike. To date, an apomictic food crop remains tantalizingly out of reach, as if on some other frequency. Researchers are not discouraged. Year after year, new findings shed light on the secrets of apomixis and seemingly bring this goal closer, although progress can be frustratingly slow and unpredictable. But the availability of technologies such as 'targeted gene editing' or 'laser capture microdissection', leading scientists argue, will advance our understanding. And one day, they claim, this will 'ultimately lead to the installation of apomixis in crop plants' (Conner and Ozias-Akins 2017:29)—with profound and radical consequences.

This narrative traces one actualization of this quest—a pioneering and at times controversial public sector and CGIAR research programme, the ORSTOM-CIMMYT Apomixis Project (OCAPo), and its metamorphosis into the ambitious 'ApoCORN' public-private partnership (PPP).[1] With goals worthy of the Apollo programme of the 1960s—but with nothing like the funding—a small team of French scientists, between 1994 and 2009, sought to revolutionize human food crops. For some, their aspirations point the way forward for agricultural biotechnology in the face of the climate emergency—in theory, apomictic wheat or maize could usher in a new era in global food security. Here I take a historical anthropological perspective on the vicissitudes of apomixis research, exploring how and why the OCAPo and ApoCORN took specific cultural, political economic and, importantly, *temporal* forms, and the impact on their development and success. To these ends, I take inspiration from social scientific and philosophical debates on temporal emergence, posthumanism and agricultural biotechnology, to make the case that a temporally nuanced approach helps anthropologists conceptualize both frontier research projects, as well as the actions of the plants themselves, and might be useful for others as well. How time was thematized and increasingly disciplined across these conjoined frontier research projects was central to their development and an anthropological perspective illuminates this. And within the projects' temporally emergent regimes for producing knowledge, the actions and differential temporalities of plants, their vegetal becomings and their generative politics were of central importance.[2] As a result, I also take the plants involved, as conceived by scientists and breeders, and in anthrophilosophical terms, as a spur for reflection and conceptual innovation.

What is apomixis? Apomixis is the ability of certain plants to reproduce asexually—or self-clone—through their seeds. It exists chiefly in 'wild' plants, but a small group of scientists across the globe are working to introduce it into staple food crops. Apomixis is considered a potentially revolutionary agricultural technology which could deliver the ability to clone F1 hybrid seed—the mainstay of modern food production. To understand why, we need to know something about how plants reproduce, how humans breed plants for food, and in particular, the importance of hybrid seed—and the implications of apomixis for them. Hybrid seed is the outcome of cross-pollination between two compatible plants. It normally results from a strategic breeding process, whereby a farmer or commercial breeder selects two plants with interesting characteristics and tries to combine them in an offspring. Today, 'hybrid seed' usually refers to seed produced on a large scale for commercial ends, but farmers and plant breeders have created it informally for much longer. Hybridization between plants can also take place without human intervention and autonomous 'hybridization events' between plant varieties or related species were historically important for the development of both maize and wheat (see Glémin et al. 2019). Hybrid seed—newly created, potentially unstable—is often contrasted with 'open pollinated' seed, which stems from a plant variety whose characteristics are relatively fixed after careful breeding over a number of generations and so breeds 'true to type'.[3] In sum, hybridization is part of the repertoire of plant reproduction techniques that humanity has adapted and coproduced with plant varieties over the millennia since farming emerged in the Neolithic—and in some way, potentially, beforehand—although it has come to commercial prominence since the early twentieth century.

F1 (Filial 1) hybrid seed is a special type of hybrid seed. It is the product of cross-pollination of two distinct plants or 'cultivars'[4] with desirable characteristics developed specifically for plant breeding. Imagine a plant breeder notices particularly valuable features in two different plants of the same variety. Over a number of years, she self-pollinates each plant in isolation—eventually over several generations the seed will hopefully produce almost identical plants. When this happens, the result is known as a 'pure line'. The goal in producing F1 seed is to *combine* these desirable characteristics in the first (F1) generation of seed through a biological process known as *heterosis* or 'hybrid vigour'. When heterosis strikes, the new cultivar is empowered with unusually enhanced qualities that surpass the desirable qualities of the parent plants. Importantly, F1 hybrids do not reproduce 'true to type' and their hybrid vigour quickly dissipates. For example, the seeds of a vigorous F1 tomato plant that shoot autonomously the following year in the gardener's allotment may be spindly and bear no fruit. They are also expensive to produce, and labour intensive, as pure lines take years of cultivation and need to be carefully maintained so that F1 seed can be produced each season. The ownership of both F1 hybrids and pure lines is therefore guarded and protected, and the seed is more expensive as a result. Successful F1 hybrid crop plants are of significant commercial value, due to these enhanced qualities, and are marketed and sold to farmers, often by large seed corporations with a global reach. And the prominence of these corporations—Bayer AG (including Bayer CropScience and Monsanto), Groupe Limagrain, Yoyodyne, Inc., Pioneer Hi-Bred International (Corteva), Syngenta AG and others—owes much to the widespread adoption of F1 seed, that went hand in hand with the spread of industrial farming.

Part of the agrarian potential of apomixis lies in the genetic preservation or 'fixing' of hybrid cultivars. If we could transfer the mechanism of apomictic reproduction into crop plants, scientists theorize, we would effectively create a self-cloning technology (Spillane et al. 2004). Very few major food crops reproduce apomictically ('self-clone'). The introduction of apomixis—from the Greek ἀπό, 'away [from]' and μίξις, 'mixing'—into F1 hybrid seed would enable the cloning of these hybrids and the preservation of their unique genomes, along with the pure lines from which they are produced. Put simply, apomictic F1 hybrids and pure lines would reproduce almost identically over the generations, and this innovation would radically change the way in which hybrid seed, and new varieties of crop plants, are produced. The potential of an 'Apomixis Technology' is widely acknowledged by scientists, and research has been the subject of significant research and development (R&D) investment by the public and private sectors. Benefits and drawbacks are claimed for a range of end users, from resource-poor farmers to transnational seed corporations, depending on the kind of technology developed (Bicknell and Bicknell 1999; Conner and Ozias-Akins 2017; Spillane et al. 2004; Van Dijk et al. 2016). However, its feasibility and implications are contested and, to date, attempts to breed or engineer apomictic crop plants have been unsuccessful. But the quasi-utopian vision, with its halo of cornucopia and hard scientific basis, continues to inspire both researchers and investors.

The travails of apomixis research are a topic on which the anthropology of science and technology casts illuminating light. This narrative is a work of anthropology—but it is written in a style that opens onto other audiences such as scientists and policymakers. Anthropological perspectives are notable for their ethnographic focus on the social, political economic and material conditions under which scientific research takes place. We explore ethnographically the actions of scientists and other human actors—but also the active role of the non-human in scientific research, be that the vibrant materiality of physical substances or the 'agency' of life forms such as plants. As scientists and plant breeders both know, the non-human has a life and character of its own, and can confound and frustrate expectations and plans. This focus on performativity and the relations between researchers, plants and other, at times more powerful influences enrols another dimension—that of time. Consideration of temporality is central to both conducting and writing about frontier research, which adheres to the new like a virus seeking its host, as we will see. Finally, discussion of apomixis also requires the transposition of botanical and scientific discourses and their redescription as emergent from socio-technical practice. But the concepts of scientists are themselves of utility to anthropology. This redescription therefore comes about through a synergy and creative alliance between scientific and anthropological discourse, which actualizes this relation in anthropological concepts that necessarily expand and adapt as a result (Viveiros de Castro 2014:187–191). This challenge unfolds as the narrative progresses—but it is important to acquire a preliminary level of scientific and technical competence in apomixis and plant breeding to secure the foundations for discussion.[5] Before elaborating further on anthropological matters, let us look again, more closely this time, at F1 hybrid seed and the case for an Apomixis Technology.

F1 hybrids and the cloning of plants

F1 hybrids are the mainstay of intensive agriculture and commercial food crops. As we now know, these commercial hybrids are created through interbreeding two 'pure lines' of a crop plant such as wheat or maize, which contain valuable traits. The pure lines are stabilized by inbreeding over many generations so that a plant's DNA transmits through sexual reproduction with minimal disturbance or mixing. (Such mixing is a normal consequence of *meiosis*, the mingling of genomes that lies at the heart of sexual reproduction in plants.) When two pure lines are interbred experimentally to combine desirable characteristics, sometimes 'hybrid vigour' is the result. In such cases, the new hybrid—the F1—is endowed with unusual strengths, such as drought resistance, high yields or high uniformity in appearance and size, that render such plants suitable for agro-industrial harvesting and markets. These strengths surpass the combined qualities of the original pure lines. However, after one or two more generations of sexual repro-duction, hybrid vigour in the new F1 plant dissipates due to the mixing and mingling of its DNA. As a result, to produce F1 hybrid seed, seed corporations have to go through the breeding process again and again—crossing pure lines to produce hybrid vigour—which is time consuming and expensive (Kingsbury 2009).

Plant scientists theorize that an Apomixis Technology would enable corporations to clone F1 hybrid seed, and recent research has provided empirical proof of this principle (Bicknell and Bicknell 1999; Sailer et al. 2016). The technology would permit improvement of hybrid seed varieties for modern agricultural systems, as the cloning cap-ability would enable the fixing and retention of a wider range of hybrid seed in seed banks, and a wider variety of pure lines and other plants. These could all be used selectively for breeding. It could also facilitate the strategic adaptation and rapid production of varieties capable of supporting food security, faced with the unpredictable consequences of the climate emergency. An Apomixis Technology is also likely to sig-nificantly reduce the costs of producing hybrid seed and so increase profits for seed corporations, given that F1 hybrids are a chief source of revenue (see GRAIN 2001).[6] Some argue that it would also facilitate the containment of genetically modified (GM) transgenes. Such a technology might take the form of a spray, for example, that when applied could 'switch on' a GM plant from sexual to apomictic reproduction, or 'turn off' the apomictic capability. Similar technologies are in use for GM crops.

In the hands of global seed corporations—which dominate international markets selling F1 hybrid seed for food—an effective, controllable Apomixis Technology could generate such outcomes. But there are other models for how it might operate. An 'open-source' Apomixis Technology, some scientists argue, might enable farmers to clone and recycle F1 seed from harvested plants without loss of hybrid vigour (Jefferson 1994). At present, farmers buy F1 hybrid seed from corporations every year, at significant expense, as it cannot be produced on the farm. Theoretically, farmers armed with an apomictic tool would be liberated from this need, and could clone commercially available F1 seed for themselves—as one might pirate and share music acquired from a CD or digital download, to use a crude analogy. This would undermine corporate profits and arguably destabilize corporate hegemony over global seed markets. Farmers would probably be sued by seed corporations for saving cloned F1 seed—in the case of patented varieties, seed saving is illegal. Monsanto and other corporations have sued farmers for seed saving, yet they are also discouraged from saving due to the uncertain quality of hybrid seed which results, as the seed itself is genetically unstable (Pollock 2000). But the scale of widespread seed saving enabled by an Apomixis Technology might undermine the ability of corporations to take legal action against farmers, even in the developed world. And in the developing world, it is unlikely that corporations would actively pursue resource-poor farmers on a large scale. An open-source Apomixis Technology might consist of a GM spray or a 'natural' facultative technology modelled on the behaviour of apomixis in the wild, that farmers could transfer to other plants through existing breeding practices. An Apomixis Technology of this kind, it is claimed, would be subversive, if not revolutionary.

Apomictic F1 hybrids released as 'international public goods' (IPGs) would comprise such an 'open-source' model. IPGs comprise several 'types' of goods with relatively free (or low-cost) access, having broad applicability across international boundaries' (Harwood et al. 2006:38). They would be widely available, and could be legally reproduced, and were the end goal of the OCAPo. As a result, the technology could theoretically support crop breeding in the developing world as well. Hybrids are a mainstay of many cultivated food crops around the world, and among resource-poor farmers, they tend to be produced locally. In this scenario, resource-poor farmers could acquire apomictic hybrids as IPGs, and, after transferring the apomictic mechanism to other plants through interbreeding, could use it themselves to 'fix' new hybrids to suit niche microclimates and unlock biological diversity. They could also use the tool to preserve other local plant varieties or useful traits. This would enable greater autonomy over the use and ownership of locally produced plants, and, potentially, support food production in the face of climate change. IPG technology would also provide local plant breeders and other commercial stakeholders with a sophisticated strategic tool and facilitate access to a greater range of genetic resources. This could enable them to preserve high-yielding hybrid varieties—perhaps the result of interbreeding or hybridization between commercial crop plants and hardy local varieties, or landraces adapted to the regional growing conditions or 'ecogeography'[7]—which would otherwise dissipate across one or two generations of reproduction and be challenging to use for seed production. In turn, this would feed into a more devolved and democratic system of plant breeding and seed distribution. In sum, the consequences of an open-source apomictic breeding tool for food security in the developing world could therefore be significant. Notably, it could also be difficult for the 'gene giants' to prevent precarious farmers in the developed world from using an open-source apomictic tool to clone F1 seed. Given such implications, the form of a future Apomixis Technology, and in particular, its relationship with intellectual property rights (IPR), are of major consequence for corporate stakeholders. Not surprisingly, IPR became a focus for ApoCORN, which forged a commercial and scientific alliance of the gene giants of the seed industry, and stakeholders for the resource poor, and sought to mediate the resulting conflicts of interest.[8]

For such reasons, the creation of a viable Apomixis Technology is not necessarily favoured by all. For example, in the early 2000s several transnational seed corporations funded postdoctoral students in public sector laboratories undertaking apomixis research. A primary objective—according to an authoritative source—was to hamper open communication of research outputs and networking through the use of confidentiality clauses (pers.comm.).[9] The production of such 'strategic ignorance' by seed corporations remains an allegation, but is regarded as entirely plausible by researchers and distinguished ethical bodies such as the Nuffield Council for Bioethics—which predicted such 'preventative action' (1999:78). An Apomixis Technology could also present a considerable environmental risk. Apomictic 'super-weeds' and aggressive rogue hybrids could be generated through widespread farming of apomictic crops and transfer of the technology by open pollination into local plants, 'weeds' and land races. Environmental campaigners therefore view apomixis research with suspicion. A number of commentators also downplay the significance of an Apomixis Technology. For example, some argue that farmers in the developed world will always purchase new F1 hybrids—while farmers in the developing world who might try cloning F1 seed do not constitute a significant market, and the quality of seed would be low due to the conditions of production and storage (Pollock 2000). The precise implications of an Apomixis Technology for the agricultural bioeconomy are therefore uncertain and disputed, although their potential magnitude is accepted by most experts.

However, most importantly, a prototype is yet to appear. During the 1990s, in the United States and Mexico, there were hopes for producing both apomictic maize and pearl millet by the millennium using plant breeding technologies called interspecific or 'wide' hybridization, and introgression (Kindiger et al. 1996; Ozias-Akins et al. 1993; Éluard and Marceau 1994). This optimism echoed important Soviet investment in an apomictic maize programme during the Cold War that reached its zenith during the 1960s and 1970s, and also failed to produce viable seed (Petrov 1984). Genomics now underpins contemporary aspirations, and in the last 20 years has established dominance over the research field, which is now largely based in hi-tech laboratories, with plant breeding approaches marginalized (e.g. Spillane et al. 2004). Yet the timeline for delivery has extended, and molecular biologists view the realization of an Apomixis Technology as a medium- to long-term goal—although hopes rise for a shorter time frame whenever breakthroughs occur. A sophisticated tool is also envisaged, of a genetically modified character, with important implications for IPR, which is more suited to the needs of corporate stakeholders than the resource poor. This corresponds to a period of increasing private sector investment; constraints on the circulation of findings due to confidentiality and proprietary agreements; a decrease in public sector funding and research capacity; and periods of low morale among researchers. For some experts these developments, driven by genomics and sophisticated advances in biotechnology, inaugurated a woeful era of scientific inertia and gridlock (Carmichael 2004; Éluard 2007). That said, new findings have recently advanced the field and renewed confidence among researchers that a technology can be realized (Khanday et al. 2019; Marimuthu et al. 2011; Sailer et al. 2016). Intriguingly, this research draws on both genomics and plant breeding techniques—confounding the advocates of molecular biology who had written off plant breeding in the early 2000s, and casting the work of the OCAPo and ApoCORN in a fresh light.[10]

Given the turbulence of their track record, today's apomixis researchers are therefore well known for talking up the radical potential of their research—while remaining unsure and frustrated about its outcomes. They have distinguished antecedents. Apomixis also confounded the founding figure of genetics, Řehoř Mendel, who in the 1860s tried to utilize hawkweed (*Hieracium*) to verify his groundbreaking findings on the rules of genetic inheritance in pea plants (*Pisum*). He did not know that the strain of hawkweed he chose (*Pilosella*) is an apomict, and the results undermined his seminal findings, which went largely unrecognized in his lifetime (Bicknell et al. 2016). Some 150 years later, unruly apomicts are still disrupting research and withholding secrets, but today's setbacks are not just due to the scientific complexities of the phenomenon. Some knowledge practices for developing an Apomixis Technology became outmoded when agricultural biotechnology entered the postgenomic era in the 2000s. But other approaches are marginalized for more complex political economic and biocultural reasons. At the same time, research capacity is limited (~175 researchers), networks fragmented, and funding restricted compared with robust frontier research fields such as synthetic biology—or, during the current coronavirus SARS-CoV-2 pandemic, vaccine research. This is intriguing, as apomixis is the only known source for a technology to control plant meiosis. Meiosis enables the mixing of genomes during sexual reproduction. Breeders depend on this unpredictable mingling of DNA for the production of new cultivars, which enables selective breeding. Meiosis, one can propose, is the abstract machine that powers modern plant breeding, placing the variables of plant genomes in relation and multiplying the opportunities for genomic differentiation and the actualization of new cultivars (see Deleuze and Guattari 1988:39–74). But it has the potential to unpick and deterritorialize its hierarchies. Its control would have ecological, political economic, and biological implications of global consequence—at a time when the climate emergency threatens catastrophe (van Dijk 2008). Such 'dysfunction' in terms of the resourcing and capacity of research arguably reflects a wider lack of action on climate change among governments and stakeholders. The challenges facing researchers and advocates of an Apomixis Technology are not merely scientific but are rooted in the socio-material foundations of this frontier research field—an ethnographic zone that falls squarely within the domain of contemporary anthropology.

Anthropology and apomixis

An Apomixis Technology, many experts believe, is likely to become a reality during the course of the twenty-first century. This narrative charts, from a historical anthropological perspective, the trajectory of the OCAPo, a leading frontier research project, and its transformation into ApoCORN, a prominent PPP. At the heart of the narrative is CIMMYT, an influential agricultural research institute based in Mexico and leading member of the CGIAR. CIMMYT is known for pioneering new varieties of staple crops, and its short-stemmed wheat drove the so-called Green Revolution with consequences that affected millions across the developing world. In the 1990s, it was the site of an ambitious project to produce apomictic maize, in international partnership with the French public sector research institute ORSTOM. The narrative traces the emergence of apomixis research at ORSTOM and the development of the OCAPo from its foundation in 1994. It then examines its transformation into ApoCORN in 2000 and the subsequent development of this PPP until its dissolution in 2009. The PPP incorporated three of the world's most influential and wealthy seed giants, Pioneer Hi-Bred, Syngenta and Groupe Limagrain. It was the first such partnership by the CGIAR, generating controversy as well as providing a model for how this influential network of agricultural research centres might cooperate with seed corporations in the twenty-first century. Such partnerships are a key template for innovation across the field of agricultural biotechnology development, and frontier research more broadly. But they pose special challenges in IPR terms, particularly to stakeholders for the resource poor and advocates for international public goods such as the CGIAR, due to potential conflicts of interest. This challenge was central to the PPP's configuration and development.

This anthropological microcosm dissects a wider historical transition in seed production and agricultural research, and here lies the projects' wider significance. The OCAPo and ApoCORN, taken in alliance, spanned that watershed transition within global seed production, from mixed plant breeding operations grounded in genetics with a public sector ethos, to the contemporary era of agricultural biotechnology and genomics where PPPs are a leading format for research and biocapital the driver (see Murphy 2007:59–153; Sunder Rajan 2006). This epistemic break in agricultural research can be termed the 'molecular turn' and was accompanied by a change in the type of Apomixis Technology favoured by scientists and research managers. As one of only two research programmes to transverse this rupture, the OCAPo and its successor ApoCORN comprise a crucible for longitudinal study.[11] In terms of anthropological work on genomics and biotechnology, the narrative contributes a focus on the field of agricultural biotechnology—of central importance to food security. A historical anthropology of the 'cloning' of plants is also a counterpart to research on animal cloning (Franklin 2007). Turning to the many studies of PPPs emerging from organizational and business studies, an anthropological focus generates important ethnographic insights into governance and innovation practices, and provides a fine-grained critique of the value of PPPs for innovation. Research was funded by the UK Economic and Social Research Council (ESRC grant RES-145-28-0001), which places a priority on open access, as part of the work of the ESRC Genomics Network, and no confidentiality agreement was signed with corporations. In methodological terms, this enabled both freedoms and restrictions. Regarding sources, some of this research was commissioned by the CGIAR Deputy Executive for an advanced workshop with leading stakeholders, hosted by the Syngenta Foundation, and supported by CIMMYT. The brief was to critically assess CGIAR's relationship with the private sector and to feed into new guidelines and policy. As a result, the narrative is partly informed by internal data furnished by CIMMYT. But it is also based on independent historical anthropological, archival and ethnographic research—including ethnographic interviews, oral history, review and discussion of scientific books and articles, and anonymized conversations with selected participants and stakeholders. In terms of access, constraints existed in terms of information that could be shared due to IPR and confidentiality agreements regarding commercially sensitive data. Further disclosure by participants and stakeholders may sharpen the picture given here—which I acknowledge at this point. The narrative offers as complete an anthropological description as possible of this leading CGIAR frontier research project and PPP, given information available at the time of writing—but I invite those with fuller knowledge of their workings to extend the depiction given here. Working with such corporate and commercial boundaries while retaining intellectual independence is central to critical anthropology and social scientific debate.

The narrative illustrates the value of a synthetic conceptual framework informed by the anthropology of time, historical anthropology and science and technology studies. It traces the development of the frontier research projects concerned and contemporary apomixis research over a number of decades. But it is also preoccupied with research activities that are explicitly oriented towards forging and taming the new, and the unpredictability this enfolds. As a result, the themes of *history* and *emergence* are integral to the narrative's conceptual foundations. This dyadic focus—on historical anthropology, and the conceptualization and narration of emergence—generates a tension that is both creative and disjunctive. Over the course of the projects' duration, I critically trace how conflictive technoscientific R&D trajectories are shaped and significant avenues of frontier research go undeveloped, which I term 'sideshadows'. How are such conflicts in time in technoscientific practices and frontier research effectively conceptualized? In response, the narrative develops an ethnographic critique of the conflictive timescapes of R&D that confront stakeholders labouring within the public sector, CGIAR and PPPs to create an Apomixis Technology.[12] I synthesize historical anthropological and philosophical perspectives to conceptualize the tribulations and uncertainties of innovation in frontier research, where aspirations to produce public goods, and the recalcitrant temporalities of biomatter, clash with powerfully calibrated mechanisms for the production of 'biocapital' (Helmreich 2008; Sunder Rajan 2006). This synthesis is also of significance to the investigation of comparable bio-innovation operations, and the wider social scientific project, where relations between history and emergence are unresolved. The conceptual tools developed here therefore resonate with current debates in the anthropology of biotechnology, but also those concerning the cross-disciplinary synthesis of historical and temporal perspectives within the social sciences and contemporary historiography.[13]

In the rest of Chapter 1, I introduce the narrative's conceptual basis and explore its implications for an anthropological perspective on apomixis. In Chapter 2, I begin with the origins of the OCAPo and explore the development of genetic, plant breeding, genomic and lab-based approaches to agricultural biotechnology within the OCAPo apomictic maize programme. I locate the project in the wider field of apomixis research, tracing its links to covert Soviet apomixis projects, the French agricultural research establishment and the CGIAR network. I also assess the importance and implications of its flexible heterogeneous structure. In Chapter 3, I then trace OCAPo's transformation into ApoCORN. I critically compare the emergent genomic 'monoculture' and *dispositif* of ApoCORN with the heterocultural assemblage [*agencement*] of genetic and plant breeding research practices within the OCAPo, and explore the slow whirlwind of changes to research organization.[14] I also dissect the processual workings of the PPP, including the conflictive temporal politics of project management and co-innovation, and their implications for technology development. Ultimately, the narrative critically surveys potential research models for producing an Apomixis Technology and how these affect end users and outcomes, and considers why an Apomixis Technology remains unrealized. The results are evidently of broader significance for frontier research and the development of biotechnologies. What becomes increasingly clear are the benefits, but also the serious challenges and obstacles facing PPP collaborations, particularly for public sector partners, where these involve technologies that may subvert the interests of stakeholders and the political economic status quo (Heisey et al. 2001). The question of whether Apomixis Technology is a wise objective, and scientifically achievable, or set to linger on a permanent basis just past the threshold of human understanding, is latent in the narrative and for the reader to ponder.

Historical emergence and the sideshadows of frontier research

The frontier research that drove the OCAPo and ApoCORN projects is unstable and exploratory. On a daily basis—the repetitive screening of thousands of maize-*Tripsacum* hybrids for apomictic traits, for example, that took place at the Mexican laboratory of CIMMYT during the 1990s—it can appear repetitive and tedious. Such research is focused on generating the new but lacks a reliable methodology for success. In addition, the new it seeks is challenging to actualize and potentially disruptive and unruly in its consequences, because it contains the potential for subverting the status quo. This is research that takes risks—with its goals, its procedures and processes, its project funding and the careers of the scientists involved. Success is not guaranteed; periodic or even catastrophic failure is possible; results are often emergent and their significance is not necessarily clear and may change over time. *Le non-savoir*, or 'nonknowledge' (Bataille 2001), is inherent to such projects, where the generation of excess information, whose value is uncertain and which may consist simply of redundant data—or even just plant waste—is central to exploration.[15] For Bataille, nonknowledge has a dynamic, materialist character and is a form of 'epistemic expenditure' or play with entropic features. It is coterminous with the production of knowledge and threatens to overwhelm it. Knowledge practices are configured in different ways to nonknowledge, and frontier research is particularly reliant on, generative of and exposed to its excesses due to its openness to the emergent. The by-product of such research is often redundant data or, in the case of plant hybrids, actual material waste—unviable seedlings, fodder for compost—nonknowledge as base materiality. One thinks of the maize-*Tripsacum* hybrids scanned by researchers seeking viable apomictic hybrids, which threatened to overwhelm the knowledge practices by their sheer volume.

'Organized uncertainty' is therefore integral to frontier research (Power 2007)—the control and attempted subjection of exploratory and emergent excess to decision-making and praxis is a key axis of project management (Luhmann 1993; O'Malley 2004). Fertile and wasteful excess is also intrinsic to historic 'hybridization events' that come about autonomously, suggesting a parallel with evolutionary development. As a preliminary insight, it is from intensive engagement with such deterritorializing states of nonknowledge that knowledge and ignorance in frontier research emerge (cf. Pedersen 2017:87–91).[16] Importantly, this conflictive timescape of challenges and uncertainties can also sow the seeds of conflict within frontier research projects. And as a further tension, such projects must also contend with a wider field of scientific practice that is more confident of its procedures and intentions for engaging nonknowledge, has more clearly defined strategies for managing risk, and often competes for the same funding. Not all scientific research is configured to *le non-savoir* in the same way.

The benefits of success justify the vicissitudes, but such uncertainty and insecurity bring several important consequences. First, there is the necessity of revisiting a project's aims and objectives at more regular intervals than for less risky endeavours—driven by the failure of workstreams to deliver, and the need to assess alternative trajectories for viability. Such research therefore slips, at times frequently, into a protean form—epochal moments that I term 'protophases'. Secondly, there is the necessity to contend, sometimes regularly, with the unexpected and the disruption and opportunities this provides. Thirdly, there is pressure to talk up the prospects for success to appease and entice funders, which generates a temporal orientation to the *promissory*. And finally, the uncertainty of project goals and their heavy implications can generate significant conflicts of interest among stakeholders. Strategies for managing these challenges and risks, which aim to cultivate and direct research emergent from the realm of nonknowledge, are therefore central to research design. As we will see, they were inherent to the OCAPo, and its successor, ApoCORN.

This distinctive focus on the emergent elicits anthropological conceptualization. As Paul Rabinow (2008:3–4) writes:

> [T]here is a difference between emphasizing reproduction and emphasizing emergence. The difference holds for the subjects in the world as much as for analysts. Most of [...] [the] social sciences concentrate on how society or culture reproduces itself (and this includes many models of 'change') [...] But there are other phenomena present today that are emergent. [Such] phenomena, it follows, require a distinctive mode of approach, an array of appropriate concepts, and almost certainly, different modes of presentation.

An anthropological interest in frontier research projects geared to produce the new evidently requires description of the ethnographic zone and temporalization of its history—in our case, over a number of years. The historical conditions of cultural reproduction provide the foundation for novelty—but the aim of such projects is to reach beyond them. To grasp the orientation towards emergence in frontier research, an ethnographic focus on cultural reproduction is insufficient. Its conceptualization demands an adjacent perspective on the striving for, and production of novelty.

From a philosophical perspective, Deleuze proposes that emergence (what he terms *le devenir*, 'becoming') 'is not history; history designates only the collection of conditions, as recent as they may be, that need to be overcome in order "to become", to create something new'.[17] To engage anthropologically with cultural practices geared to emergence, one must bring their specific temporal orientation into focus and trace how it is actualized in everyday life. By contrast with studies of cultural reproduction, which seek to identify pattern, frequency and repetition, an important methodological and theoretical emphasis for engaging with emergence is therefore on 'placing oneself within the flow of the [unfolding] event in its becoming [...] to pass through each of its elements and each singularity' (Rabinow 2011:62). Rabinow (ibid.) divides these two approaches into that of the *thinker*, whose task is 'to seize an event in its becoming', and the *historian*, whose role 'is to insist on the importance of historical elements as conditioning whatever takes place'. But these distinctive roles can also be articulated transversally and coexist—and they must.[18] Such thematization of historical timespace in the making generates the need for an integral focus on the *temporal*.

L'intempestif *and temporal emergence*

Within the wider field of technoscientific and agronomic research, frontier research on apomixis is an activity that seeks the new in a particular form. This can be characterized by what Deleuze terms *l'intempestif*. *L'intempestif* is Deleuze's preferred translation of Nietzsche's German term *Unzeitgemässe*—and an improvement on the English equivalent, 'untimely'. As Rabinow (2011:60) writes:

> [T]he term captures a striving to bring something forth, something that could be actual but does not yet exist [...] [Such a discovery] reconfigures existing things and relations, and it would be inopportune in that it disrupts those existing things and relations and changes their tone, register, and directionality.

L'intempestif evokes a range of meanings—the ill-timed, the inopportune, the untimely and the disruption it can bring. A focus on the botanically new as an emergent goal and catalyst for radical change is a defining feature of the quest for an Apomixis Technology. The degree of change and disruption generated—social, environmental, biological, political economic, scientific—would vary, for example, depending on the technology's form and relationship to IPR. But although scientists speculate about what these changes would involve, their unpredictability and potential unruliness remains a defining quality. It is this unruly alliance with the new that *l'intempestif* invokes. As Deleuze states, '*l'intempestif* arrives in an era *against* its era. [It is] what comes at a present moment against the present in order to prepare [...] a future' (Deleuze 1987:19, my emphasis; cf. Rabinow 2011:59–62).

As a first temporal observation, then, frontier research can be productively conceptualized as oriented towards an open field of temporal emergence. The concept of temporal emergence conjures an image of historical becoming as open, contingent, performative, unpredictable and ontologically novel (Deleuze 2004; Hodges 2008). It also enables the conceptualization of frontier research in 'posthuman' terms as interacting with realms of non-human agency—in this case, plants as they pertain to apomictic reproduction (Pickering 1995). However, and importantly, emergence is not treated here as a foundational temporality of life itself, but an anthrophilosophical concept that evokes a specific *ethnographic mode of being in time*. As an image of thought, it is a tool of descriptive equivalence not a category of metaphysical realism—becoming is not *real* but its conceptual utility *is*. Other temporal dispositions also characterize the ethnographic zone, it must be emphasized, as we see later in the narrative. But scientific and economic preoccupation with *l'intempestif* foregrounds this temporal mode within the timescapes of both the OCAPo and ApoCORN, and underpins their orientation towards uncertainty and the unruly excess of nonknowledge.

The sideshadows of frontier research

From an anthropological perspective, frontier research is therefore characterized by a practical orientation towards historical becoming, framed as an open field of temporal emergence. The scope of these emergent possibilities, it is clear, may dilate at particular junctures or moments of crisis. At other times—of setback, failure, or stasis, for example—it may contract. Anthropological conceptualization requires consideration of the social and material factors influencing how emergent possibilities are actualized, choices made and research trajectories followed (Williams and Edge 1996:866–867; see also Latour

2007; Stone 2014). Given the protean form of frontier research, it also involves exploration of why alternative counterfactual research pathways are ignored, sidelined, marginalized, blocked, obviated, forgotten or go unrecognized through activities of knowledge production or the ruinous agency of nonknowledge. Nonknowledge in this sense is not simply the 'unknown', those generative 'lies of exclusion' of which Nietzsche (1989[1873]:246–257) wrote and by virtue of whose exclusion from discourse the known ('truth') comes into focus—or *l'impensé*, the 'unthought' that overshadows the boundaries of discourse (Foucault 1966). It is an excess of becoming from which knowledge and ignorance arise, and that, in the case of frontier research in particular, threatens to overwhelm. This protean, differentiating zone of play, dissolution and creativity riddles knowledge practices, is internal or virtual to them, and is partly generative of them. In this regard, *le non-savoir* outreaches and encompasses both knowledge and ignorance, and even when successfully displaced, remains latent at their core. For once knowledge and ignorance are mutually actualized, they remain in emergent tension and their relation may be wholly reconfigured in the future.[19]

These real-world counterfactuals to actualized research trajectories can be conceptualized as a research field's 'sideshadows'. For Morson (1994:118, 120–121; emphasis in original):

> Sideshadowing relies on a concept of time as a *field of possibilities*. Each moment has a set of possible events (though by no means every conceivable event) that could take place in it. From this field a single event emerges [...] This is a simultaneity not *in* time but *of* times: we do not see contradictory actualities, but one possibility that was actualized and, at the same moment, another that could have been but was not [...] Time exists not just 'phenotypically' but also 'genotypically'.

The concept of sideshadowing rests on the insight that 'the temporal world consists not just of actualities and impossibilities but also of *real though unactualized possibilities*' (Morson 1994:119, my emphasis). Sideshadows can be of ephemeral or durable consequence. Their potential for actualization can be tied to a contingent event, or endure in timespace. In this way, historical timespace can be grasped as a set of differential events and unfoldings, a garden of forking paths incorporating the actual *and* its sideshadows. For an example of a technological sideshadow, consider the electric car, which competed with the petrol car in the early twentieth century and was considered superior on several fronts, including lack of emissions. For complex reasons including the vested interests of petrochemical corporations, its development and production was sidelined. But given its recent prominence, it now counts as a 'durable' sideshadow of motor vehicle technology. It follows that to conceptualize frontier research on Apomixis Technology in terms of temporal emergence, one must trace and conceive of time's *virtual fullness*. The OCAPo and ApoCORN can be viewed as preoccupied with the emergent sideshadows of Apomixis Technology development—with facilitating their actualization and potentially preventing it, or at times as simply awash in the state of latent nonknowledge that infused the projects.

Morson's figure of sideshadowing offers a route for the anthropological conceptualization of frontier research in apomixis, but this can be extended through synthesis with Deleuze's concept of the 'virtual'. Novelty and recurrence, for Deleuze, arise through the actualization of the virtual rather than the realization of 'possible' events, as with Morson. The possible, Deleuze (1988, 2004) argues, exists only in retrospect. While the new may be possible before it exists in that there is nothing to prevent its occurring, this does not mean that its actual occurrence is concretely foreseeable, for as a 'possibility' it only exists once it has occurred. One must instead see the virtual as productive of the actual. It is the tendency that produces the actual, a de jure rather than de facto principle of differentiation which is never actualized.

Deleuze's theorization deepens the conceptual underpinnings of the notion of sideshadows and connects it to STS theories in which emergence is thematized (Latour 2007; Pickering 1995). Morson's 'field of possibilities' can be recast as a plane of virtualities that have the potential to be actualized within specific historical events. The concrete form they take is immanent and emergent to the ethnographic zone in which they are actualized. The temporal politics at stake—frontier research, or PPPs focused on technology development—may involve strategic contest over which virtual trajectory is actualized, and importantly, how and when this takes place. Timing is everything, not only in terms of the intentions of agents but also with respect to the ways in which contingent events regulate the trajectories that can be actualized, and anchor the emergent character of their actualization (Deleuze 2007:231; Williams and Edge 1996:873–875). This struggle is intensified during moments of crisis, or *protophases*, where research practice becomes uncertain and is deterritorialized to a significant degree as the excess of nonknowledge—also emergent from the virtual plane of sideshadows—threatens to overwhelm or destroy it. That said, not every twist and turn in research is reducible to such intentions, and the shaping of knowledge and production of lacunae and ignorance may have humdrum or accidental motivations—such as the actions of researchers whose horizons are affectively shaped by working routines, bureaucracy, pragmatic concerns about their careers or affective enthusiasms for new technologies and outcomes. Bovensiepen (2020:502–504) identifies the affective 'banality' that characterizes bureaucratic praxis among oil company employees, and the *Gedankenlosigkeit* ('lack of thought') it generates, as this calibrates situational knowledge and ignorance about environmental impacts. Such knowledge and ignorance is not strategically produced, but still enables 'marginalisation of uncomfortable knowledge through a process of reasoning, routines and collective solidarity'. Similar telescoping of vision within frontier research, for related reasons, can generate important consequences in technoscientific and human terms. Likewise, it may also be a by-product of the unruly actions of plants—or triggered by the surfeit and deterritorializing forces of nonknowledge itself. There are many contingencies and reasons why sideshadows remain virtual, or are actualized, and where appropriate, these are also considered.

Transversal relations and historical emergence

Building on these insights, the narrative focuses on the period 1994–2009. However, the goal is not to redescribe the emergent contingency of history as a causal historicist narrative. Through a selective focus on key events, the objective is to adumbrate their virtual fullness, conflicts and becomings, and conceptualize their interrelations. An ongoing focus is the actualization of the apomixis research field's sideshadows—an anthropology of emergence. But this necessarily incorporates a sibling—anthropological documentation of the historical conditions from which the new, *l'intempestif*, emerged. If 'history designates only the collection of conditions [...] that need to be overcome in order "to become", to create something new' (Deleuze quoted in Rabinow 2011:62), the narrative's twin objectives generate a paradox. How can historical anthropology and the anthropology of emergence coexist?

To accomplish this, I pursue a *transversal* approach. Transversality in a literal sense signifies 'cross-cutting'. For Deleuze, it refers more precisely to the assembly of 'heterogeneous components under a unifying viewpoint [or narrator] [...] [which] draws a line of communication through heterogeneous pieces and fragments that refuse to belong to a whole, that are parts of different wholes' (Parr 2010:291–292). In theoretical and, importantly, political terms, 'the function of transversals is to assemble multiplicities, yet in such a way that the differences among entities are not effaced but intensified' (Bogue 2016:2). Where a historical anthropological perspective comes to the fore, this therefore takes place within a *transversal* ethnographic narrative incorporating different temporal frames and evocations that seek to capture the historical event in its becoming. This plane of historicity, assembling multiple perspectives, is what Lundy (2012:63) terms the 'differential element of "the between" [...] a "history/becoming" that is simultaneously more than static history and less than eternal becoming'. In this way, historical and emergent anthropological perspectives on historical timespace in the making coexist as a 'creative composite' (ibid.:183) that transversally allies historical, theoretical and ethnographic sources on the past and transforms them through narration.[20]

Redescribing 'process'

Finally, to execute this temporal manoeuvre, a key trope in anthropological thinking must be displaced. Since the 1980s, 'process' has emerged as a central and governing trope in the Western social-scientific imaginary (e.g. Smith 1982; Wolf 1982). The dominant notion of timespace that underwrites contemporary anthropology is couched in the 'processual idiom', linked to a shift in the temporality of thought from static, atemporal frames to approaches grounded in the ontological assumption that human (and non-human) life exists in 'time', 'flow' or 'flux'.[21] However, the cultural foundations for this shift are opaque, nor has the conceptual meaning of this influential trope been extensively debated (Hodges 2008:400–403; Lyman 2007). This 'processual idiom' operates 'insofar as all such totalizations abstract from the concrete multiplicity of differential times co-existing in the global "now" a single differential [...] through which to mark the time of the present' (Osborne 1995:28). In this sense, it anchors anthropological thought in a monological temporal outlook that obscures as well as enlightens.

'Process' has become such an integral cog in the doxa of social science that it is easy to forget that it is a socio-historical *concept*, a cultural figure with which to frame action and conjure 'time' (Arendt 1958:230–236). In its common form, anthropologists use the trope of 'process' to construct the transcendent temporal unity of cultural practices. These are normally cast as bounded or open-ended 'processes' incorporating change but that exhibit a transcendent, systematic set of linkages 'over time', to be documented ethnographically and elucidated theoretically.[22] Simultaneously, process can invoke a diachronic, spatialized temporal foundation for study that views cultural practice as part of a 'manifold, a totality of interconnected processes' (Wolf 1982:3; see also Bourdieu 1977; Fabian 1983; Nicholson and Dupré 2018). Such uses are not necessarily teleological and can acknowledge emergence. And processes are sometimes said to coexist and operate at different tempos. But it is uncontroversial to state that anthropologists rarely define or clarify this constitutive processual ontology. The concept of process is used to construct a relation in which past-present-future are conjoined in an epochal moment (see Agamben 2000:227). It is also used to invoke the soul of time, a spatialized, riverine flow or flux in and through which processes unfold—that cousin to linear or homogeneous empty time (Hodges 2008:399–400). As a result, process operates as a foundational concept, providing a constructed epochal moment or temporal 'clearing' that serves as a frame for anthropological discussion.

Following Arendt (1958), this conceptual dominance reflects the influential and varied role of processes in industrial societies. Most significantly, it echoes the subjection of raw materials and people to *procedures* of production. For Arendt, such procedures instrumentalize social relations and 'things' into means which are subsumed into end products and their correlates in profit. In this sense, 'process' is a key socio-material archetype for modern social organization—a development echoed in the increasing visibility of processual idioms in Western scientific, political, cultural and historical discourses from the eighteenth century onwards.[23] Such heterogeneity also suggests that processualism is a polythetic category of cultural practices with academic and wider variants. If Arendt is ambivalent about the value of processualism (Arendt 1958:305–309), Jameson (1998:169–170) is critical, arguing that it is open to:

> other doubts and suspicions, particularly in a society whose current economic rhythms perpetuate and thrive on permanent change: capital accumulation, investment and realization, the dissolution of stable firms and jobs into a flux of new and provisional entities, awash in structural unemployment, its cultural infrastructure committed to permanent revolution in fashion and to the imperative to generate new kinds of commodities, [or] in deeper crises [...] wholly new production technologies.

For Jameson, the hegemony of processualism is darkly symptomatic of neoliberal political economic practices. This view is echoed in Koselleck's discussion of the temporality of 'modernity', marked by the ideology that history and time are an incessant movement or process to which every historical object and actor is subordinated, and by a hegemonic processualism operating at the level of social organization (1985:255–275; see also Blackwell and Seabrook 1993).What is clear is that the monological character of processual idioms can obscure those conflicts in time that become apparent with use of a differential, non-spatialized temporal idiom (e.g. Adam 1998; Gurvitch 1964).

Rather than taking 'processes' as a temporal foundation, I conceptualize life as an immanent timescape comprised of multiple conflictive temporalities, of different tempos and durations. 'Process' is therefore relegated to a trope and component of cultural practices—although it evidently remains in the anthropological toolbox. Where I discuss processes, therefore, I focus on their temporal and epochal construction, discursive agency, material basis and conceptual manifestation, in other words on how processes are 'achieved' (Whitehead 1979:208–218). This is important as, in ethnographic terms, ApoCORN invoked processualism at the level of social organization, scientific idiom and the genomics *dispositif* in an echo of Arendt's critique. In this sense, the conceptual premises informing frontier research in Apomixis are similar to those active in the discourse of anthropology itself. The conjuncture between the two idioms—social scientific and scientific—therefore compels the subversion of anthropology's temporalizing concepts, which are displaced, coproduced and redescribed as a result.[24] In this sense, anthropology is conceived as an immanent relational practice generative of novelty in anthropological knowledge and speculative philosophy.[25] It is not merely engaged with the 'conceptual worlds' of others, but affective, experiential, non-human and ethnographic ones (Biehl and Locke 2017). As Viveiros de Castro (2015:27, emphasis removed) writes, anthropology's 'constitutive role is not that of explaining the world of the other, but rather of multiplying our world'. This is not just a result of engagement with the generative conceptual principles of scientists' worlds but also with the ethnographic plane of immanence—life itself that outreaches the conceptual and whose blocs of sensation and becoming anthropologists invoke through literary-ethnographic discourse.[26] The 'anthropological relation' thereby produces a synergy and disjunctive alliance with the ethnographic Other—an uncanny echo, in some ways, of interspecific hybridization but on the human level.

Anthropology and the politics of plants

What do these insights hold in store for an anthropology of the OCAPo and ApoCORN? Sometimes, the virtual sideshadows of technology development are as significant as the actualizations that emerge. For example, genomic advocates widely dismiss the sideshadows of apomixis research—the plant breeding *dispositif* that dominated the field prior to the 1990s—as misconceived. Yet there is no evidence that when used in tandem with genomics, it would not deliver an apomictic tool—although its form may subvert corporate dominance within the agricultural bioeconomy (Carmichael, 2004; Dressel et al. 2001; GRAIN, 2001). Other factors are evidently in play. Indeed, the work of Raphael Mercier and colleagues at INRA, whose recent research has advanced the field, demonstrates the value of a hybrid, public sector approach, even if this remains in the minority (see Marimuthu et al. 2011).[27] In theory, the use of plant breeding techniques in frontier research projects may even constitute a strength, echoing as it does at times the emergent trajectory of evolutionary events.

The politics of which virtual research trajectories are actualized, and which remain sideshadows, is tied here to the conflictive temporalities of knowledge practices within the OCAPo and ApoCORN. It is most visible in two key protophases—epochal moments of uncertainty and rupture—during the 2000–2004 period. I interrogate how specific research trajectories were actualized in relation to competing sideshadows, and explore cui bono.[28] I also assess whether certain research sideshadows were deliberately and strategically blocked given their challenge to corporate interests. But one must also take account of other reasons for why sideshadows go unactualized, which, as noted above, is not necessarily the result of conscious strategy, and can also be the result of the actions of non-humans—plants. In anthropological terms, ApoCORN was also an ethnographic zone in which processual temporalities were conspicuous and increasingly enabling. There was a movement from the flexible assemblage of the OCAPo, driven by the practice of interspecific hybridization, in which emergence is thematized—towards an ApoCORN disciplinary programme[29] and *dispositif* enabled by plant genomics and the interests of stakeholders including corporate partners, where controlled productivity is a focus. This was predicated on a marked shift in the social organization of knowledge practices from the OCAPo assemblage to ApoCORN *dispositif*. I pursue this historical movement away from accommodation of the temporal modalities of plant reproduction through interspecific hybridization—which incorporated disciplinary scientific technologies as a component of heterogeneous knowledge practices—towards disciplining of the temporalities of project work and plant reproduction through the use of processual technologies of power and plant genomics. Processualism must therefore become a target for ethnographic critique rather than a master trope of anthropological conceptualization. In this way, *disciplining time* became a focus for research activity within ApoCORN—which in turn, entangled the project with wider disciplinary programmes and processual regimes of truth within the global bioeconomy (Arendt 1958).[30]

Finally, at the heart of this study are the plants themselves—hybrid assemblages [*agencements*] of maize (*Zea mays*) and its apomictic 'wild' relative, Eastern gamagrass (*Tripsacum dactyloides*), which asserted a disruptive agency in both projects. These unruly hybrids—emergent from technoscientific practices and relationally joined with humans—shaped, undermined and thwarted research plans, disciplinary governance and innovation strategies. They confounded human ambition and desire for knowledge and success, on multiple levels. They also acted as agents of deterritorialization and nonknowledge at the centre of both projects. Landecker (2007:232–233) proposes that

> the operationalization of biological time is a dominant characteristic of the interactions of humans and cells in technical environments over the last fifty years […] In short, living matter is now assumed to be stuff that can be stopped and started at will.

The objective of the OCAPo and ApoCORN was to short-circuit the actualization of the sexual sideshadows of maize plants, driven by meiosis, and install an apomictic reproductive system in its place. In turn, this would lead to seeds that contained a clone of maternal DNA, and from this 'virtual Idea' a new plant would be actualized, both the same as and different to the maternal plant (cf. Deleuze 2004:168–221). But not all living matter is cooperative, as Landecker suggests. The role of botanical unruliness in the OCAPo and ApoCORN constitutes a unifying theme of this narrative, tracing how plants relationally enable and subvert the efforts of scientists to accommodate and discipline plant-time, vegetal becomings and reproductive processes. This outreaches an ethnographic focus on scientists and anthropology's fixation on *anthropos* to encompass plants themselves—apprehending them relationally via the people joined to them (see Strathern 2017). Some write of anthropology as philosophy that includes the people(s) (Ingold 2014). One can similarly conceive of an anthropological counterpart to the philosophy of plants and plant biology—anthropology with the plants in, an 'anthrobotany', perhaps (see Marder 2013; Myers 2015).

In this sense, 'cloned' plant DNA can be conceptualized as a virtual Idea from which the new plant is actualized and, in turn, shaped relationally by the circumstances of its actualization (Deleuze 2004:280). The apomictic progeny is the same as (in terms of DNA), and different to both the maternal plant and other progeny. Although scientists informally refer to apomicts as 'clones', apomictic self-cloning via seeds is complex, and clearly differs from popular understandings of 'cloning' as a process of identical replication.[31] It also differs from other forms of plant reproduction by 'cloning', such as that exemplified by rhizomes like yams, taro, cassava and so on, which have been cultivated by humans for thousands of years. The philosophical concept of the 'rhizome' is widely influential in contemporary social theory (Deleuze and Guattari 1988). One might think this provides an effective tool for conceptualizing apomixis and the activities of researchers working on it. By contrast, the actions of apomictic plants problematize its botanical correlation and equivalence in intriguing ways (see also Strathern 2017:24). The apomict is notable for subverting the transcendental filiation of seeds and installing the immanent in its place in a form that is neither rhizomatic nor filiative (the 'clone' takes the place of the progeny). Its apomictic subversion of sexual reproduction (filiation) via seeds therefore renders it both different and repetitive—an immanent differentiation (through actualization) of the same virtual Idea (cloned maternal DNA) in the form of a progeny (not a rhizome). At times it reverts to filiation (facultative), and at times it opts for pure immanence (obligate).

In this way, Haudricourt's (1962) influential distinction between immanent (rhizomatic, Eastern) and transcendental (filiative, Western) agricultural regimes—a formative influence on both Deleuze and Guattari's thinking and the discipline of ethnobotany[32]—is subverted and transmuted. For Haudricourt, dominant forms of agriculture and the plants around which they are constructed have a central role in the historical formation of the constitutive differences between Eastern and Western civilizations, as they compel dominant types of action that resonate across cultural domains. The East, for example, is said to predominantly cultivate rhizomes—generating archetypal practices leading to a nomadic, immanent 'culture of clones' (Haudricourt 1964:95) and the development, for example, of religions and philosophies focused on 'immanence'.[33] The image of the West is that of the sower or shepherd, which in Deleuze and Guattari's (1988) terms is associated with philosophies of transcendence, and likewise resonates in their distinction between royal and nomadic sciences. A related genre of suggestive 'botanical determinism' is also present among apomixis researchers, as will become apparent. Anthropologists have problematized Haudricourt's thesis through ethnographic erosion of the immanence-transcendence binary (see Rainbird 2001). Apomixis does the same for the reproductive dualism on which his thesis is based. But it remains suggestive of ways in which 'non-binary' apomictic crops—neither filiative nor immanent but a disjunctive synthesis of both—might influence global agriculture.

Finally, taking this one stage further, one can also ask, what kind of conceptual figure does apomixis elicit? If one were to extract a concept from apomixis, in the vein of the plant-philosophy or becoming-plant of Deleuze and Guattari, rather than the rhizome's differentiation and repetition, one might conceive of an assemblage of the intensive drive to differentiate associated with the immanence of the new, allied in a disjunctive synthesis to the tendency to extend, transcend and repeat symptomatic of the genealogical imagination. Or a counternatural alliance of context (transcendence) and novelty (immanence) that connects historical and emergent conceptions of, and perspectives on life, human and non-human.[34] For rhizome, read *apomict*. The result is a novel image of thought in which differentiation and emergence are placed on the same plane as the desire to transcend—a conception that implies equivalence, although *not* an equivalence in utility. One outcome is that the tendency to explicitly, or implicitly, ground anthropology in the master trope of ontological differentiation is upended by the insight that differentiation is itself, in the last instance, a Western conceptual technology (Hodges 2008). If this move affirms 'the equivalence, in principle, of anthropological and indigenous discourse' (Viveiros de Castro 2014:189), then, it does so conscious that the 'metaphysical ground' for such an insight must itself be equivalent.[35] With such insights in hand, we turn to consider the historical trajectory of apomixis research and the emergence of the OCAPo.

APOMIXIS AND ITS RESEARCH FIELD—The term 'apomixis' designates a mode of plant reproduction whereby plants produce seed asexually which contains a replica of the maternal genome. It is relatively common, with examples among the *Poaceae* (e.g. maize or pearl millet), *Rosaceae* (e.g. crab apples) and *Asteraceae* or *Compositae* families (e.g. dandelion). Very few agricultural crops have an apomictic breeding system, the main exceptions being tropical and subtropical fruit trees, for example, mangosteen, mango and *Citrus*; apple; orchids; and tropical forage grasses, for example, *Brachiaria*, *Dichanthium*, *Panicum*, *Pennisetum* and *Poa* (Asker and Jerling 1992; Bashaw and Hanna 1990). The phenomenon was noted by botanists during the 1830s at the Royal Botanic Gardens, Kew, when John Smith (1841) noticed that three female specimens of *Alchornea ilicifolia* (native holly) that arrived from Australia in 1829 produced seeds without the presence of male plants. The term was probably first used in 1906 by the German botanist Hans Winkler, who also coined the term 'genome' in 1920.[36]

There are two main types of apomixis as classified in contemporary technoscientific discourse, *gametophytic* and *sporophytic*, based on the site of origin and subsequent development route of the cell that generates the embryo. Both are theorized to occur as a result of 'disruption' or 'deregulation' of the sexual reproductive pathway, a processualist framing that emerged in the United States in the 1960s and now grounds scientific conceptualizations worldwide. Sporophytic apomixis, also known as 'adventitious embryony', is less common and typically occurs in *Citrus*. It takes place when one or more cloned embryos emerge directly from the nucellus (central part) or integument (skin) of the ovule (the structure that contains the female reproductive cells) at an early stage within the normal sexual reproductive pathway. The apomictic embryo(s) may then invade the sexual embryo sac that develops alongside it, and compete with the sexual embryo within for the nutrients it contains.

Gametophytic apomixis gives rise to megagametophytes, or 'embryo sacs', not embryos. There are two main types, 'apospory' and 'diplospory', depending on the stage in the sexual reproductive cycle in which the apomictic event initiates. In apospory, an embryo sac arises from a cell in the nucellus. It then develops into an embryo without fertilization. More than one embryo sac may develop, resulting in multiple embryos being produced, or what is termed 'polyembryony', as with sporophytic apomixis. In diplospory, an embryo sac emerges from the 'megaspore mother cell', a structure that develops at a later stage in sexual reproduction than apospory (and sporophytic apomixis). Both embryos arise as a result of 'parthenogenesis', or 'virgin birth', as neither is the result of fertilization, and other apomictic variations are also possible, such as 'apogamy' and 'mixed apomixis'.

Notably, plants capable of reproducing by apomixis do not always do so—this is an important variable with multiple implications. An 'obligate apomict' *always* reproduces apomictically (although some scientists dispute whether this is ever the case in the final analysis). A 'facultative apomict' *sometimes* produces a sexually reproducing specimen, and, importantly, the frequency varies. Most natural apomicts are facultative, which—scientists agree—facilitates the maintenance of biological diversity. Apomictic processes also vary between species, as does their genomic basis. And although apomictic plants 'self-clone' the maternal genome within their seeds, the expression of their DNA during growth and development leads to instances of differentiation or becoming between individual plants—due to epigenetic and environmental factors, for example—while mutations can arise within DNA replication (Schmidt 2020; Ferreira de Carvalho et al. 2016).

Finally, turning to the apomixis research field, this is a well-integrated, relatively small global network, given the potential importance of an Apomixis Technology, whose exact numbers are difficult to estimate as corporate research also takes place confidentially. There are probably about 150 molecular biologists worldwide focused on apomixis research, working on a range of research foci and model plant systems, and publishing largely in English. These draw on a range of scientific approaches to Apomixis Technology development, which remain theory based. Notably, the numbers of scientists working openly on Apomixis Technology are more limited, and many apomixis researchers claim to work purely on the phenomenon itself. Leading public sector laboratories are located in Canada, France, Germany, Mexico, Switzerland and the United States, with important research also taking place in Australia, India and in other countries. Private sector research is not easily quantifiable, but leading seed corporations have pursued important long-term programmes, as have crop improvement companies such as KeyGene.

In the postgenomic era, researchers roughly divide into those studying apomixis as a deregulation of sexual reproduction, via analysis of non-apomictic model systems such as *Arabidopsis thaliana*, and those studying natural apomicts and aiming to understand and reconstruct apomixis through 'reverse engineering'. Plant breeding was previously the foundation for research into apomixis, and while it is no longer the dominant research paradigm—as we will see—some important contemporary laboratories combine research programmes in genomics with practice-based research using advanced plant breeding research paradigms and techniques. A leading example is the Department of Chromosome Biology at the Max Planck Institute for Plant Breeding Research. They have produced intriguing and innovative results.[37] In the case of the apomict that provides a key focus for the OCAPo and ApoCORN, *Tripsacum dactyloides* (Eastern gamagrass) reproduces through 'facultative diplospory'. It is a member of the *Poaceae* family and a wild relative of maize (*Zea mays*).The intention was originally to transfer its apomictic mechanism into maize plants through 'interspecific hybridization', an advanced plant breeding technique. The introduction of genomics and the creation of ApoCORN modified these goals.

Notes

1 The acronym CIMMYT stands for Centro Internacional de Mejoramiento de Maíz y Trigo (International Maize and Wheat Improvement Centre). CGIAR, or the Consultative Group for International Agricultural Research, is a global stakeholder for resource-poor farmers and a leader in the translation of scientific research into resource-poor agriculture, partly through production of 'international public goods' (Harwood et al. 2006). ORSTOM, funded by the French state, stands for Office de la recherche scientifique et technique d'outre-mer, or Overseas Scientific and Technical Research Institute. In 1999, ORSTOM became IRD, Institut de recherche pour le développement (Research Institute for Development).

2 Marder (2013:95–107) discusses the differential (hetero-)temporalities of plant life.

3 Open pollination also refers to the process by which such seed is reproduced— through pollination by insects, birds or other animals, or the wind, for example. Sometimes by humans. When such a plant is pollinated by another plant of the same variety, or self-pollinates, the offspring is very similar to its parent—'true to type'. Other techniques for seed production and seed saving evidently exist—too many to mention here.

4 A cultivar is a plant selected and valued for its desirable characteristics. Such a plant is normally created through human cultivation.

5 Further technical detail, of greater complexity, is located in the text box at the end of chapter 1.

6 Savings were previously estimated at 2.5 billion $US per annum for hybrid rice (McMeniman and Lubulwa 1997). The figure is likely to be significantly higher today.

7 A landrace is 'a cultivated, genetically heterogeneous variety that has evolved in a certain ecogeographical area and is therefore adapted to the edaphic and climatic conditions and to its traditional management and uses [...] [and] have evolved and may continue evolving, using conventional or modern breeding techniques, in traditional or new agricultural environments within a defined ecogeographical area and under the influence of the local human culture' (Casañas et al. 2017:1).

8 See Jefferson 1994 and Curtis et al. 2004 for an overview, and Spillane et al. 2004 and van Dijk et al. 2016 for an assessment of the benefits of an Apomixis Technology. Van Dijk and van Damme 2000 and Grain 2001 address the environmental risks and GM issues.

9 Where names and dates are not given for personal communications, they have been anonymized.

10 Navashin and Karpechenko grasped the potential of apomixis for agriculture in the 1930s (Asker and Jerling 1992:263). For overviews of routes to an Apomixis Technology, see Carmichael 2004; Conner and Ozias-Akins 2017; Dressel, Carmichael and Éluard 2001; Spillane, Curtis and Grossniklaus 2004; van Dijk 2008.

11 The other programme was based at the United States Department of Agriculture (USDA) (Bashaw and Hanna, 1990). Since the 1970s, Wayne Hanna and his team had worked at the USDA to transfer apomixis from *Pennisetum squamulatum*, a wild grass, to pearl millet.

12 For Adam (1998:11, 56), '[a] timescape perspective conceives of the conflictual interpenetration of temporalities as an interactive and mutually constituting whole [.] Where other scapes such as landscapes mark the spatial features of past and present activities and interactions [...] timescapes emphasize their rhythmicities, their timings and tempos, their changes and contingencies'.

13 See Bear 2014; Bryant and Knight 2019; Kleinberg 2012; Lambek 2018; Palmié and Stewart 2016.

14 For Deleuze and Guattari, 'assemblage [*agencement*] is a topological concept that designates [...] the productive result of the intersection of two [or more] open systems [...] its properties are emergent [and] only discernible as a result of the intersection of both such systems' (Marcus and Saka 2006:103). The term designates flexible social or non-human composites that give a 'structural quality to a contingent object of heterogeneous relations' (ibid.:104). For Rabinow (2003:56), assemblages are 'comparatively effervescent' and *disordered*, as distinct from stable systems of knowledge practice covered by the term *dispositif* (apparatus), although the boundary is ambiguous (see also Rabinow 2011:122–124). I use assemblage in contradistinction to *dispositif*, which focuses on the control, regulation and generation of situated knowledges through disciplinary practices (Foucault 1980:194; cf. Legg 2011:130).

15 Bataille's (2001) writings on *le non-savoir* ('nonknowledge') are aligned to his work on 'general economy', expenditure and excess, or the 'accursed share'. Rather than signifying a state of 'ignorance', nonknowledge synthesizes elements of his concepts of 'inner experience', 'sovereignty' and 'base materialism' (Bataille 2014 [1943]; 1985 [1930]). The conjunction between the conceptual domain of scientific discourse and the unruly material domain of plants in frontier research exemplifies Bataille's base materialist configuration in an intriguing way (see Lerman 2015:117–137; Noys 1998). And if nonknowledge threatens to overwhelm and ruin such projects, at times they produce unexpected and startling advances—to recall William Blake (2011[1793]:66), '[t]he road of excess leads to the palace of wisdom'. In Deleuze and Guattari's terms, nonknowledge is characterized by 'lines of flight', 'deterritorializations' and proximity to the 'plane of immanence' (Deleuze and Guattari 1988, 1994; cf. Viveiros de Castro 2015:22).

16 Deterritorialization is 'the movement by which something escapes or departs from a given Territory [...] [It] is always bound up with correlative processes of reterritorialization, which does not mean returning to the original territory but rather the ways in which deterritorialized elements recombine and enter into new relations' (Parr 2005:70; see Deleuze and Guattari 1988:508–510).

17 I use Rabinow's (2011:62) translation from an interview with Deleuze by Toni Negri. The French reads: 'il y a deux manières de considérer l'événement, l'une qui consiste à passer le long de l'événement, à en recueillir l'effectuation dans l'histoire, mais l'autre à remonter l'événement, à s'installer en lui comme dans un devenir, à rajeunir et à vieillir en lui tout à la fois, à passer par toutes ses composantes ou singularités. Le devenir n'est pas de l'histoire; l'histoire désigne seulement l'ensemble des conditions si récentes soient-elles, dont on se détourne pour "devenir", c'est-à-dire pour créer quelque chose de nouveau. C'est exactement ce que Nietzsche appelle l'Intempestif. Mai 68 a été la manifestation, l'irruption d'un devenir à l'état pur' (Deleuze 2007:231). See also Deleuze 1995:170–171; Deleuze and Guattari 1994:111.

18 See Lundy 2012 for a related conception that he terms 'historiophilosophy'.

19 Pedersen (2017:88) posits a conception of 'ignorance' as an 'infrastructure' from which 'certainties and lacks of certainties are undergirded and mutually brought into being'. This conception resembles Bataille's (2001) formulation of nonknowledge but is ethnographic in focus, referring to 'the way in which lack of knowledge is built into the texture of life in peri-urban Ulaanbaatar' (Pedersen 2017:89).

20 See also Hodges 2019; Lambek 2018; Palmié and Stewart 2016.

21 Lyman (2007:220–224) reviews anthropological usages of process prior to the 1980s, illustrating how it did not occupy the foundational place it does today.

22 For Rescher (2000:22): 'A process is an actual or possible occurrence that consists of an integrated series of connected developments [...] that are systematically linked to one another either causally or functionally [...] Processes develop over time: any particular [...] process combines existence in the present with tentacles that reach into the past and future (2000: 22)'. He identifies three key characteristics: 1.) A process is a complex of occurrences—a unity of distinct stages or phases. 2.) This complex of occurrences has a certain temporal coherence and integrity, and processes accordingly have a temporal dimension. 3.) A process has a structure, a formal generic patterning of occurrence, through which its temporal phases exhibit a fixed format (2000:24).

23 Another key processual idiom in Western discourse can be traced to the pre-Socratics, notably Heraclitus, and was influential in shaping the social sciences through phenomenological philosophy (e.g. Heidegger 1993a; Schutz 1967). By contrast, philosophers such as Whitehead (1979) propose a radical concept of 'process', closer to the immanent philosophy of Spinoza and Deleuze, arguing that any occasion belonging to a 'process'

is an incidence ('concrescence') of novelty or form of 'birth' with no transcendent frame.

24 Compare Franklin 2013 on anthropological transformations resulting from
    research on in vitro fertilization.

25 See De Beistegui (2010) for philosophical discussion of immanence, and
    Altieri (1979) for an assessment of its importance in contemporary poetry.
    Viveiros de Castro (2014) arguably elaborates a project for an immanent
    anthropology.

26 Anthropological conceptualizing is distinct from philosophy, in that it is
    enmeshed in life through its immersion in the ethnographic. Anthropological
    discourse—whether focused on the contemporary or the historical—is
    a composite that outreaches the conceptual to engage perceptual and
    affective dimensions of life through aesthetic techniques of evocation, in a
    hybrid of conceptual and aesthetic discourse: 'By means of the material, the
    aim of art is to wrest the percept from perceptions of objects and the states
    of a perceiving subject, to wrest the affect from affections as the transition
    from one state to another: to extract a bloc of sensations, a pure being of
    sensations' (Deleuze and Guattari 1994:167).

27 INRA is the Institut national de la recherche agronomique (National
    Institute of Agricultural Research), and is a public sector organization
    funded by the French state. It was founded in 1946, and is one of the world's
    leading agricultural research institutes, operating under the authority of the
    French Ministries of Research and of Agriculture. In 2019, Mercier was
    invited to direct a new Department of Chromosome Biology at the Max
    Planck Institute for Plant Breeding Research, in which molecular research
    and plant breeding research paradigms coexist.

28 Through reflection on why certain trajectories were followed, and others
    were not, the production of scientific ignorance comes into focus as a
    cultural practice (Proctor and Schiebinger 2008; cf. High et al. 2012;
    Stone 2014).

29 '"Disciplines" in Foucauldian terms [are] those micromechanisms
    [*dispositifs*] of power whereby individuals [and plants] are molded to serve
    the needs of power' (Ransom 1997:59). Foucault defines such practices as
    central to the reshaping of power relations in Western societies over the past
    two centuries, and a sociological feature of wider modernity. Disciplinary
    programmes operationalize them, and 'define a domain of social reality to
    be turned into an object of rational knowledge, intervened in and made
    functional' (Gledhill 1994:148) which is implemented through technologies
    of power (appropriately designed practices), according to contingent strat-
    egies. Foucault (1977) documented the genealogy of disciplinary power in
    relation to humans, but it also applies to the relationship between humans
    and their environment, including plants, and, importantly, *time*.

30 See Ransom 1997. I use 'regime' in Foucault's (1977) conception of a dom-
    inant set of cultural practices productive of a discourse that assumes the
    doxic guise of truth (cf. Hodges 2019:392–393).

31 Franklin (2003:71) draws a similar conclusion for animals, while noting the imprecision of the term 'cloning'. For important biological reviews, see Koltunow and Grossniklaus 2003; Léon-Martínez and Vielle-Calzada 2019; Éluard 2000.

32 The term first appears in French in Haudricourt and Hédin's *L'Homme et les plantes cultivées* (1943).

33 See Ferret 2012:116 for discussion of Haudricourt's binary contrast between 'direct positive action' (West, seed, sheepherding, transcendence) and 'indirect negative action' (East, cultivation of rhizome, immanence).

34 I deploy a concept of alliance defined as 'a movement that deterritorializes the two [or more] terms of the relation it creates, by extracting from the relations defining them in order to link them via a new "partial connection" […] Alliance ceases to designate an institution—a structure—and begins to function as a power and potential: a becoming' (Viveiros de Castro 2014:160, 166). There are also cultural and political forms of alliance—institutions, for example, that require ethnographic description and anthropological conceptualization, or indigenous concepts that parallel and differ from anthropological ones and elicit their transformation. But in terms of the *relation* it generates, and as a starting-point for conceptualization, every alliance is posited here as a becoming—just as 'every becoming is an alliance' (ibid.:164).

35 Which is not the same as saying everything is relative. See Strathern (1988:17–18) for a well-known image of thought seemingly grounded in equivalence. Stasch's (2009) ethnography of an indigenous 'philosophy of difference' among the Korowai of West Papua, Indonesia, renders Western ontologies of difference equivalent by implication.

36 Hans Winkler (1877–1945) was Professor of Botany at the University of Hamburg. In analyzing how *Wikstroemia indica* (tie bush), an apomictic shrub found in Southeast Asia and elsewhere, produced its seeds, he proposed 'the term Apomixis, which was formed in analogy with Amphimixis [sexual reproduction, literally 'mixing both'], and which can therefore be defined as: replacement of absent sexual reproduction by another, asexual reproductive process' (Winkler 1906:253, my translation).

37 See: www.mpipz.mpg.de/chromosome-biology, accessed 28 August 2020.

2 The quest for apomictic maize at CIMMYT

Apomixis research at ORSTOM

An hour's drive outside Mexico City, across a landscape seeded with crops of variegated Mexican maize, wheat and beans, stands the station of the Centro Internacional de Mejoramiento de Maíz y Trigo, or CIMMYT as it is known. The contrast between the flatlands of the Mexican plateau, the city of Texcoco rotating with centrifugal cool about the ghost of Nezahualcóyotl, and the modern research institute set among sculpted fields of wheat and maize is striking and perhaps deceptive. CIMMYT grew out of an initiative of the Rockefeller Foundation and Mexican government in the 1940s. It was fired by an almost religious belief in the value of plant breeding for food security, inspired by US achievements with crop improvement during the Great Depression. The programme quickly developed international networks of researchers to test experimental crop varieties, which saw Mexico self-sufficient in wheat production by the late 1950s. Under the scientific leadership of the lauded American agronomist Norman Ernest Borlaug (1914–2009), during the 1960s it went on to release short-stemmed, high-yielding varieties of hybrid wheat that were adopted by India, Pakistan and other developing countries. Dr Borlaug was awarded the 1970 Nobel Peace Prize for his work. Famously, if controversially, these innovative crops, in tandem with agricultural 'modernization' programmes promoting monocultures and industrial farming, endorsed by Borlaug and CIMMYT, precipitated the 'Green Revolution'.

CIMMYT's agricultural research programme is well known for pioneering improved varieties of staple crops and (Haudricourt might argue) seeding a major international shift in farming techniques, social organization and religious values. It is also credited (counterfactually) with saving the developing world from a punishing famine during the 1970s. The Centre assumed its current institutional identity in 1966, and has developed a diverse funding base of governments, development agencies, philanthropic organizations and, through PPPs, can also draw on partnerships with transnational corporations. As an early member of the CGIAR, the coalition of international agricultural research centres working for poverty alleviation through scientific innovation and collective action, it zeroed on developing new wheat and maize crops for resource-poor farmers.[1] CIMMYT's mission is to channel (*translate*) the most up-to-date scientific knowledge and innovation to some of the poorest people on the planet. This has morphed hand in hand with an openness to technological emergence—that comes with its own set of political economic drivers. In 1990, it opened an Applied Biotechnology Centre to supplement its established plant breeding practices, which kick-started engagement with genetic and genomic technologies over subsequent years (see Murphy 2007:86–87). More recently, in 2013 it received donations of ~$23 million from the billionaires Carlos Slim and Bill Gates to grow its technological base with new biotechnology labs, doubling its research capacity, and is also a vocal supporter of the importance of GM crops for increasing productivity and meeting the needs of the rising global population. This approach—controversial for some—speaks to its historic willingness to push the boundaries of science in the quest to improve food security and, more recently, to tackle the challenges of the climate emergency.[2] There is no doubt this line of flight saved countless lives over the past 50 years—while the world was watching moon shots … checking its phone.[3] Largely reterritorialized by agribusiness … Syngenta, Limagrain, Pioneer Hi-Bred etc. … *ex nunc* by apomixis.

In the vein of this distinguished living tradition of agricultural innovation, during the 1990s CIMMYT hosted a world-leading frontier research project that aspired to create the first apomictic cereal crop. The OCAPo was founded with the objective of breeding apomictic maize, through 'wide hybridization' of *Zea mays* with an apomictic relative, *Tripsacum dactyloides* or 'Eastern gamagrass' (Éluard and Marceau 1994). Researchers aimed to integrate this established plant breeding technique with technologies such as large population screening, flow cytometry and genomic in situ hybridization, to create apomictic maize. The project was emergent from the intersection of CIMMYT's maize programme with the agro-technological culture of the French overseas research and development institute, ORSTOM.[4] It was tuned into historic wavelengths in apomixis research underwritten by advances in, and synergies between biotechnology and plant breeding, and related technological breakthroughs (Kloppenburg 2004; Murphy 2007). More specifically, it reflected CIMMYT's ongoing dependence on funding from France and other Western governments, whose policies demanded engagement with hard science and new biotechnologies (IRD 1998; Jasanoff 2005). The OCAPo can be anthropologically conceptualized as an emergent research assemblage [*agencement*] (Rabinow 2011:122–124) whose flexible structure enabled heterogeneous alliances of differing lab and field-based research techniques, and human and plant actors, and their transformation under scientific leadership. In this regard, managed disorder, manifested through the creative disjunction of nonknowledge, was as central to its operational focus on the emergent as order (Legg 2011:128–129). But it was catalyzed by one researcher's entrepreneurial vision. *Cut to*—an impromptu round table on apomixis convened by Dmitri Fedorovich Petrov, the distinguished long-term director of the Soviet Maize-*Tripsacum* project (and USSR Master of Sports in Chess) at the 14th International Congress of Genetics in Moscow, many years before—

This landmark congress, held in the mild Moscow August of 1978, was entitled 'Genetics and Human Welfare', a prescient title, given the aspirations that would mark the apomixis research field. It was the first major international gathering of apomixis researchers, and an epochal juncture in the historical ethnographic zone. But it was an awkward affair for that very reason, and at the daytime panel on apomixis— Thomas Éluard would remember years later—debate had stalled. The evening round table brought together 30 scientists from the West and the Soviet Union to discuss their findings with the help of a Polish translator, who was located at the last minute. The informal atmosphere thawed the day's awkwardness—after years of working in isolation, an exchange of ideas flared among the weary researchers. Among those present were early pioneers of maize-*Tripsacum* hybridization, who knew little about each other's work, including Petrov himself and Dr Jack Harlan, the botanist from the University of Illinois who trained under the celebrated evolutionary biologist, G. Ledyard Stebbins.[5] Harlan was conducting research into maize-*Tripsacum* hybrids at the Illinois Agricultural Experiment Station (AIES), but with no serious thought of creating a commercial apomictic crop. Petrov, on the other hand, grasped the potential of introgressing apomixis into maize in the 1940s, after witnessing the consequences of food shortages in the Soviet Union before the war, including the devastating *Holodomor* and Soviet famines of 1932–1933. He had been unsuccessful in his hopes to outwit *l'intempestif* and forge apomictic maize, but his work, undermined by Stalin's post-war purge of Soviet genetics, recovered to make strong growth in the 1960s and 1970s, and he built up substantial results. Also present was Dr Éluard, a 33-year-old French scientist, who was invited to co-chair the daytime Congress session by the septuagenarian Russian veteran. Years later, when I spoke with Éluard of that heady encounter, he recalled that so little was known about Soviet apomixis research in the West, they'd thought Dmitri Petrov was younger than it turned out.[6]

On the occasions I met with Éluard, he spoke with passion about the potential value of *l'apomixie* for crops and for solving the challenges of global food security—and about sailing on the perilous *golfe du Lion*, where he had his long-term home. He was wiry, energetic, with a blast of white hair, and still passionate about the potential of apomixis for agriculture. From one year to the next, as I visit, his enthusiasm does not waver, and in his memoir published years later—the same engaged voice (Éluard 2020). At the time of the Congress, Éluard had already begun to consider how apomixis might bear fruit for plant breeding. ORSTOM had begun work on apomixis more than ten years earlier, when Paul Debord selected the tropical forage grass *Panicum maximum* (Guinea grass) for a breeding study—*because it was known as an obligate apomict*, Éluard recalled, *and theoretically could not be bred*. Debord, an anarchist spirit in the eyes of his colleagues, was attracted to the conundrum of a plant with a huge morphological diversity that was rumoured to reproduce through self-cloning. 'If there's diversity, there's sexuality', he reputedly said (Éluard 2020). His curiosity was further aroused when he returned from a field trip to East Africa with what appeared to be 250 apomictic samples, only to discover during lab analysis that one of these plants was in reality a sexual diploid—an apparently natural 'error' or anomaly, he initially thought, faced with the impenetrable.[7]

This early encounter with *l'intempestif*—or as Éluard conceived of it, with an individualistic conceptual frame, Debord's uncanny good fortune and bold, scientific personality—gave rise to the inaugural project of ORSTOM's new plant genetics group. For it was Debord's chance encounter with the sexual diploid plant, and decision to grasp this emergent opportunity by revising his thinking and taking a chance with an experimental study, that successfully advanced research at the historical juncture where ORSTOM was open to supporting such work. But Debord's response to the event also constitutes in methodological terms a concrete exemplification of the relational, performative 'dance of agency' that shapes scientific research. As Pickering (1995:22; cf. Bennett 2010) writes:

> The dance of agency [...] takes the form of a dialectic of resistance and accommodation, where resistance denotes the failure to achieve an intended capture of agency in practice, and accommodation an active human strategy of response to resistance, which can

include revisions to goals and intentions as well as to the material form in question and to the human frame of gestures and social relations that surround it.

The fact that Guinea grass exceeded or 'resisted' Debord's expectations sparked the 'dialectic of resistance and accommodation' which characterizes this temporally emergent practice—in Pickering's conceptualization—as in response, Debord revised his thinking and developed novel strategies and practices of intellectual enquiry to pursue the 'dance'.[8] It also demonstrates how historical and emergent perspectives can be transversally allied in historical anthropological description. If 'history designates only the collection of conditions that need to be overcome in order [...] to create something new', the concept of the dance enables us, in ethnographic imagery, to 'seize an event in its becoming' (Deleuze, qtd. in Rabinow 2011:62). The resulting transversal narrative of historical conditioning and emergence—historical events coexisting with, and preceding Debord's encounter, and the emergent encounter with *Panicum maximum*—generates a disjunctive synthesis of history and becoming. To these ends, I draw on historical and ethnographic evidence—from interviews, archival and documentary sources—and social scientific concepts, to install the book's narrative within the flow of the event in its becoming and capture an impression of how the sideshadows of apomixis research can be actualized.

Debord had a background in agronomy, and a practical and curious mind. He undertook a field trip to Kenya to collect specimens of *Panicum maximum* as early as 1967. Over the coming years, he developed a wide interest in hybridization events across neighbouring species and their evolutionary import. The group he founded was known as Biologie et Amelioration des Plantes Utiles (Biology and Improvement of Useful Plants), a name which had a Zen-like simplicity to it. In time it cultivated a unique school of students tuned to the vitality and importance of interspecific hybridization, with a scientific ethos marked by independent and original thought. In time, this gifted fieldworker became an influential figure in French botany and apomixis research, and a professor at the University of Paris. And Debord's associates saw an echo of the value he sought in a chance encounter with genetic exceptions within a field of identical plants—the emergent 'error' that can trigger a sea change in plant breeding or even evolutionary terms—in the colleagues and students he selected to work with him (Mounolou and Sarr 1990). Debord accepted Thomas Éluard as a doctoral student in 1972, and tasked him to analyze the first *Panicum maximum* hybrid progeny. Debord, with his appetite for seeking the sideshadows thrown up by meiosis on his treks through offbeat fields of nonknowledge, was Éluard's route into the apomixis research field. And you can imagine that Éluard's break was due to his resonance with the singular poetics of the apomictic *intempestif* that infused Debord's image of thought. He was one of those independent-minded research students, an exception to the norm, Debord judged, needed to study the apomicts. It was a logical development, although seemingly uncanny in retrospect. Botanical archetypes, as Haudricourt (1962) posited, burrow through the cognitive soils of cultural subjects like rhizomes in a bed of earth—when you are a botanist or a philosopher, perhaps. Half a century later, Éluard (2020) would dedicate his memoir to Debord in a gesture of intellectual filiation.

Panicum maximum is a 'facultative' apomict. This means that in certain populations, a small number of plants (circa 8 per cent) reproduce sexually (Éluard 1982). Most natural apomicts are facultative, which, scientists theorize, facilitates the maintenance of biological diversity. Without sexuality, of course, Debord would have been unable to breed *Panicum*, but its facultative capability was not well understood at the time. The discovery prompted further trips to the field by Debord and his team, which located additional sexual specimens. As the programme developed, *Panicum* hybrids were gradually bred and developed— ORSTOM eventually released them on international markets. Several of Debord's offspring continue to be planted across the developing world today, of which the most prevalent is ORSTOM T58, rereleased by EMBRAPA in the late 1980s as Tanzania-1.[9] It was an inspiring story as far as plant breeding goes—whose success was not lost on Éluard. Debord demonstrated that apomictic plants could be the foundation for a breeding programme—as paradoxical as that once sounded. The *Panicum maximum* hybrids were the emergent product of a dance of agency between plant geneticists, plant breeders and the plants themselves, to identify the principal actors. This was no normal example of plant breeding innovation and commercial reproduction, but a case of *l'intempestif* tamed through temporally emergent scientific and breeding practice. In conceptual terms, the outcome was actualization of this important sideshadow of the apomixis research field. Debord's project, in which Éluard played an important role, also demonstrated the viability of a research heteroculture that synthesized cultures of genetics and plant breeding. And importantly, for this narrative, it delivered practical lessons in how apomixis could operate as a breeding tool for commercial ends, but also when used independently by farmers.

For anthropological purposes, it also elicits the relational transformation of anthrophilosophical concepts. Success with *Panicum maximum* demonstrates the role and value of a deterritorializing base materialism (Bataille 1985) operative within the *facultative* apomictic mechanism—with nonknowledge reconceived as emergent from vibrant matter (Bennett 2010). The self-cloning *Panicum* apomict—a *causa immanens* of filiation that produces by remaining in itself—normally deregulates sexual reproduction to install itself through an act of reterritorialization as the apomictic progeny. But to unlock this closed circuit and create a breeding tool, the sexual act must return and filiation by meiosis temporarily restart. Meanwhile, agronomists selected for their archetypal resonance with *l'intempestif* are engaged to develop relations with hybrid progeny, in languages of genetics and plant breeding which mediate this redescription, to direct the procedure. And although the plants consent to practical submission in the guise of an apomictic cultivar, ultimately they subvert and outrun the scientific (and anthropological) concepts as their functioning remains mysterious. Protean chaos and relational surprise thereby enter the anthropological conceptualization of technology, as in turn, it too engages the field of nonknowledge.

And how 'identical' is the apomictic clone? In conceptual terms, cloned DNA operates as a sideshadow or virtual Idea from which the new plant is actualized—the progeny is both the same (virtually) as the maternal plant, *and* different when actualized from seeds into a new cultivar within a specific ecogeographic zone. This insight was also the basis for technoscientific conceptualization among ORSTOM scientists of the time, which, importantly, was couched in the Western-inflected agential discourse of genetics. But due to its facultative capability, the 'cloning' mechanism is ultimately of utility for both plants and humans precisely through its unpredictable *openness* to emergent lines of flight and sexual sideshadows. Successful apomictic replication of maternal DNA therefore defers to *historical conditioning* via actualization during the plant's real-world growth and differentiation—and to *becoming* at the level of the virtual Idea (DNA itself) through the meiosis of facultative sexual reproduction (~8 per cent). History *with* becoming ... Emergent sexuality is also central to the evolutionary viability of the plant, and, importantly, the breeding system. Emergence *with* filiation ... Both apomictic and sexual pathways are combined and strategically directed within the *Panicum* plant breeding assemblage—which relies on historical conditioning *with* emergence (the historian *with* the thinker) ... Which produces, we can conclude, an equivalence in principle (relational dyads but *not* dualisms). Finally, such achievements at ORSTOM in breeding Guinea grass, in alliance with Debord's enthusiasm for the importance of cross-species hybridization, both facilitated Éluard's initial appointment and cultivated his appetite for ambitious apomictic breeding programmes. It is a series of resonances and becomings that Haudricourt (1962) would have recognized.

Redescribing wide hybridization

Let us step back from historical events at this point, to review the OCAPo's technical foundations in more detail. 'Interspecific' or 'wide' hybridization, also known as 'wide-crossing', involves cross-fertilizing two plants of distinct but related genera. Its objective is the 'introgression' of a trait via specialist technoscientific and plant breeding practices from one genus to another—usually a 'wild' relative to a domesticated plant—for example, apomixis to maize from an undomesticated relative. The targeted trait for the OCAPo, of course, was apomixis. 'Introgression' refers to the specific process of transferring the gene or gene cluster from one gene pool to another. The technique often depends on creating numerous experimental crosses with the aim of locating the objective among the hybrid progenies—in an echo of what Debord hoped to find in his fields. Advanced technologies facilitate this quest. The resulting field of nonknowledge is potentially unruly, and populated with the creative chaos of plant biomatter. The objective is counternatural selection by technoscience.

When a promising interspecific or wide hybrid with the trait is discovered, it is 'backcrossed' (interbred) with a plant from the target genus to remove unwanted hybrid traits and undesirable characteristics. This procedure typically takes several generations of plant reproduction, which may be numbered Backcross (BC) 1, BC2 and so on, in relation to the original hybrid. In the case of maize-*Tripsacum* hybrids, after crossing maize with gamagrass, the wide hybrid would be backcrossed with the maize parent to remove other gamagrass features and possible mutations, with the aim of producing 'pure' stable apomictic maize—a 'pure line' of a distinctive kind. One strength of the technique is that it can be deployed by professional plant breeders, scientists *and*, in a more rudimentary sense, farmers. Resource-poor farmers in the developing world, for example, might wide-cross commercial crops with landraces adapted to local microclimates to enhance resilience, and refine the outcome through backcrossing. Some botanists—such as Paul Debord—scour far-off fields seeking impromptu interspecific (or intraspecific) hybrids whose value is yet to be determined, as a basis for such work. However, without technoscientific support, the technique is ultimately dependent on the ability of the plant breeder to encourage and harness fortuitous emergence and *l'intempestif*.

The technique of interspecific hybridization is modelled on key evolutionary operations—historic wide hybridization events have contributed to the development of many crops. It therefore mimics a 'naturally' occurring phenomenon, whereby plants from related genera 'accidentally' cross and reproduce to create interspecific hybrids. *Triticum aestivum* (common bread wheat), for example, emerged from wide hybridization events over extended timescales, many occurring before domestication (Glémin et al. 2019). Throughout the late twentieth century, scientists also debated whether maize emerged from a wide hybridization event involving *Tripsacum* and an ancient maize relative, or was domesticated by humans directly from teosinte (*Zea mays* subsp. *Parviglumis*), a Mexican grass. The latter was eventually confirmed using new DNA technologies and archaeological evidence (Stitzer and Ross-Ibarra 2018). A novel form of vegetal historiography also emerges, as new technoscientific practices ally with the progenies of interspecific hybridization to burrow into the hidden histories of plant DNA and temporalize the virtual hybridization events of deep time. The virtual Idea of DNA—the past that coexists with (pre-exists) the present that does not cease to pass and impels the future—is temporalized as a historiographical and cartographic template for both past and future knowledge and becomings (Hodges 2008:410–412). History (*historicism*) with emergence.

Ultimately, interspecific hybridization is both a 'naturally' occurring phenomenon and a method that farmers are likely to have used since the dawn of agriculture. As a plant breeding technique, it is combined with genetic and molecular technologies, and is a sophisticated tool. In anthropological terms, it is a technique where temporal emergence is methodologically *thematized*, as the unpredictable encounters of alien genomes during meiosis are placed at the heart of technical practice. Concretely, it exemplifies how, throughout the history of plant evolution, the role of historical conditioning and natural selection operative through sexual reproduction, *even when* this incorporates meiosis and the genomically new, is displaced by the order of pure becoming— *un devenir à l'état pur*—at the moment when a transversal alliance is generated between species.[10] This may sometimes take place through mutation—but when considering Apomixis Technology, we are particularly interested in the role of humans, whose relational participation is joined (disjunctively) to the emergent hybrid (differentiating) becoming-species. In this sense, interspecific hybridization can thereby be conceived as a practical improvisation on the concept of disjunctive synthesis, 'a relational mode that does not have similarity or identity as its (formal or final) cause, but divergence or distance' (Viveiros de

Castro 2014:112)—although its material (plant) form is clearly of another order.

Taking these insights forward, scientists confidently divide maize and *Tripsacum* into two distinct species, but posit that they diverged from a common ancestor < 1.2 million years ago (Gault et al. 2018). This is the conceptual ground, of course, for the OCAPo interspecific hybridization project. The proposed hybrid alliance is not founded, therefore, on the amalgamation of two pure forms into a third, but rests in the first instance on the historic relation between the two species, which are different and related at the genomic level. For example, they share prominent morphological features, such as the spikelet arrangement and cupulate fruitcase so characteristic of corn. Such perceptions of prior relations are, of course, the result of quite recent acts of botanical and technoscientific redescription, although the indigenous 'ethnobotanists' of Central America have no doubt perceived similarities between maize and gamagrass for significantly longer. Interspecific hybridization, in this sense, can be conceptualized as the eliciting and reanimation of these 'prior' virtual relations through a novel conjunction with humans—forming a chimeric human-non-human assemblage from which plant-becoming steps off. These relations are not the basis for a new conjoined identity, however, but one starting point, among others, for the emergence of the new apomictic maize plant that would diverge or disjoin rather than unify.

But what does 'hybridization' signify, in this sense? For scientists, these are two distinct species that are related, which forms a rationale for their recombination (hybridization). That said, their understanding will differ perspectivally depending on the genetic or genomic paradigm invoked. From a Latourian (2007) perspective, by contrast, the resulting 'hybrid' emerges precisely through the formation and stabilization of this network of social and natural actors, in a very different conceptualization. Anthrobotanically speaking, however, these are not two pure and ideal plant forms that meld, or an emergent Latourian hybrid, but can be conceptualized as a set of dispersed relations on the virtual plane with recombinatory potential existing within the heterogeneity of contemporary speciation. Wide hybridization as a plant breeding technique activates this potential through the intervention of humans, who act as catalyst or agent, placing these relations into disjunctive orbit and igniting interspecific alliance to produce (they hope) the new. On the way, hybridization as a concept begins to lose its traction, in anthropological terms at least, and the workings of emergence become an increasing focus. '[B]ecoming is a movement', writes Viveiros de Castro (2014:160), 'that deterritorializes the two terms of the relation it creates, by extracting them from the relations defining them in order to link them via a new "partial connection"'. And by contrast with the scientists' image of dissolved and recomposed boundaries, we indeed find the virtual potential of reactivation—tracing how one set of inter-specific relations retreats as a residual, coexisting set comes to the fore and is temporalized from the virtual past to catalyze a new connection, and ultimately *l'intempestif* itself. Interspecific hybridization, one can propose, in this regard, can be conceptualized as becoming with a his-torical edge—dyad not dualism—as the perspectives of the historian (conditioning) and thinker (becoming) once again seem to combine (cf. Rabinow 2011:62).

Such interspecific hybridization events can result in a sterile assem-blage [*agencement*] from which no plant can reproduce—and in the case of maize-*Tripsacum* hybrids, they often did. But when the out-come is a functional apomictic progeny—or a sexual one—the result is an instance of pure becoming (Deleuze 2007:231). It is in this dis-junctive synthesis and interspecific alliance, and its relational cap-acity to generate lines of flight, that the potency of wide hybridization lies, and which the OCAPo team sought to channel. It exists on a different frequency from the 'banal' everyday deterritorializations and reterritorializations of intraspecies meiosis or plant-*Gedankenlosigkeit*. I adapt here Bovensiepen's (2020) conception of the affective 'banality' that can produce situational ignorance—modifying in her turn Arendt, who reformed Heidegger—that burrowing rhizome of thought—to cap-ture the repetition through differentiation of plant 'habits' of reproduc-tion (*l'ensemble des conditions*) that need to be overcome (*dont on se détourne*) in order to create something new. Plants and humans both, within a knowledge-producing assemblage such as the OCAPo, take a role in generating pathways of 'ignorance'—as does nonknowledge, those *real* fields—and it is clear that vegetal relationality and vegetive compliance and vegetative resistance and vegetational distraction and indifference take proliferous forms (cf. Deleuze 2007:231; Marder 2013). In technical terms (as acknowledged by researchers) it was also possible that what science did not understand about apomixis and the challenge of knitting together two plant species—the plants might pragmatically 'work out for themselves'. In redescribing interspecific hybridization, then, this account does not simply replicate or 'explain' ethnograph-ically what it signified for the OCAPo and its team. It repurposes this technique and its cultural understanding within the project in equivocal alliance with anthropological discourse—the discourse of this book—in a gesture of interdisciplinary hybridization and disjunctive synthesis between plant breeding, scientific discourse, anthrobotany and plant philosophy, for anthropological ends (Lebner 2017:3–5; Viveiros de Castro 2014:41).

Genetics and apomixis

The work of Petrov and Paul Debord was a precedent and inspiration, but in terms of epistemic practices—sprawling, many levelled—the contemporary epoch of apomixis research commenced for real in the early 1980s when two important, and closely related, genetic paradigms of apomictic reproduction gained acceptance. Both were embedded in public sector plant breeding research contexts, knowledge practices and experimental systems. Taken together, their emerging hegemony underwrote the fusion of apomixis research with the quest for an Apomixis Technology that has remained a trystero of the field. Early intimations of the first paradigm emerged from Jack Harlan's research (Harlan et al. 1964). Importantly, Harlan proposed that apomixis and sexuality were not independent but relational and potentially interactive modes of reproduction. This countered theories that apomixis contained a distinct reproductive system, and suggested that the sexual pathway might be enabling of, and ultimately subverted by, apomictic events and processes. In the field of nonknowledge that encompassed apomixis at the time, Harlan's theorem was a scientific exemplar of that individualistic sexual diploid that Debord sought, with its fecund intensities and emergent potential for redescription of the boundaries of knowledge and ignorance. The proposal that apomixis comprises a *deregulation* of sexual reproduction in space and time—perhaps resembling an error or malfunction,[11] or a derritorialization or line of flight in anthrobotanical terms (Deleuze and Guattari 1988:508–510)—is now a leading view in the postgenomic ethnographic zone. It also facilitates the epistemic integration of apomixis researchers with scientists working on sexual reproduction in plants—a substantially larger and well-resourced field of research (e.g. Koltunow and Grossniklaus 2003). It is notable that the plant genomics currently enabling this prominence is markedly distinct in theory, practice and objectives from the post-war fusion of classical genetics and plant breeding practised by Harlan—despite the similarity in their views of how apomictic reproduction unfolds. Harlan's theorem turned out to be rhizomatic news that stays news, travelling way underground and up through the postboxes of contemporary apomictic researchers. This scientific reconceptualization triggered by plant encounters took on new life as it transited across scientific paradigms into the cultural-conceptual processualism of genomics technoscience—before a transition rendered its temporally emergent sideshadow intelligible in anthropological terms.

The second paradigm was predicated to an extent on Harlan's 'deregulation thesis'. It proposed that apomixis, particularly the variant of apomictic reproduction known as 'apospory', has a relatively 'simple' genetic regulation.[12] Arguments emerged in the late 1970s, in Switzerland and France, in unrelated work by Gian Nögler and Thomas Éluard. They drew on Nögler's research with *Ranunculus auricomus* (buttercup), and Éluard's work with *Panicum maximum*. In both cases, technically speaking, Nögler and Éluard proposed that one pair of alleles controlled the switch to apospory from sexuality (Éluard 2000). Other apomixis researchers were following ORSTOM's research on *Panicum*, and some had extrapolated that the apomictic mechanism might be composite enough to introgress into crops. With the emergence of Éluard's and Nögler's research, a well-developed scientific rationale for this hypothesis was now in place, grounded in the knowledge practices and culture of genetics, where the identities of plant 'genes' echo Western core symbols of individualism and agency rather than the diffuse and capillary discourse of genomics with its epigenetic tendrils. Such bounded 'gene-individuals' could transplant and resume their raison d'être and activity elsewhere with greater ease, like moving apartment blocks. And if the genetic trigger for apomixis was relatively compact, as state-of-the-art research then suggested, its transfer into crop plants by interspecific hybridization and introgression now appeared feasible, even though limited progress had been made at the time in analyzing the genetic regulation of apomixis across different plant species. The hypothesis that a compact genetic actor was also present in *Tripsacum* and, potentially, could be transferred to create apomictic maize was seductive. It was a neat credible recipe for actualizing an apomictic sideshadow ... and hardly tempestuous. And sure enough, it soon emerged as a key premise on which the OCAPo was founded—and was indispensable to its conception. However, importantly, and unlike Harlan's influential 'deregulation thesis', it did not outlive the impending genomics revolution of the millennium.

Such theoretical insights and developments were enabled by wider technoscientific discourses. The broader hegemony of plant breeding within apomixis research at the time derived from the historical association of genetics with plant breeding paradigms and the study of inheritance, which dated back to Řehoř Mendel's nineteenth-century experiments (Schwartz 2008). It also derived from the practical orientation of leading scientific researchers towards achieving an Apomixis

Technology, which received new momentum from the Moscow Congress of 1978. The field of plant breeding had absorbed innovations from molecular genetics since the 1950s (Kloppenburg 2004). These allied with long-term epistemic practices rooted in Mendelian and quantitative genetics. A process of historical becoming was at work, in which genetic knowledge practices were translated from lab-based experimental systems to plant breeding research cultures, and vice versa, in a form of epistemic hybridization or disjunctive differentiation that advanced the field (Rheinberger 1997:179–181). In turn, this invigorated genetic knowledge, in tandem with the emerging biotechnologies which amplified its development at tremendous pressure, fuelled the increasingly globalized political economy of agro-industrial research during the 1980s.

Éluard's research on the genetics of apomixis had initially developed during the 1970s. It matured within ORSTOM experimental research systems where classical and molecular genetics were practised and tested in plant breeding for the applied goals of international development. It was also influenced by Debord's ethos and valorization of the opportunities presented by interspecific hybridization. In the midst of this wider technoscientific assemblage, and with the help of emerging and important networks among apomixis researchers—which he played a leading role in developing—by the early 1990s Éluard's duplex of conceptual-breeding practices achieved prominence over Nögler's pure research. Yet at a global level, the heyday of public sector institutions that supported the historical fusion of genetics paradigms with plant breeding for the purposes of international development and applied ends was drawing to a close. Genomics and the private sector were in the ascendant, which would have important implications for the OCAPo (Bonneuil and Thomas 2009; Murphy 2007:126–153).

Prompted by success in breeding Guinea grass for Debord, Éluard initially drafted a proposal to wide-cross pearl millet (*Pennisetum glaucum*) with apomictic relatives of the *Pennisetum* genus. 'Results were showing that very likely apomixis was, you know, simple enough to be a potential tool', he would recall. But on his return from Moscow in 1978, he penned a report on the Congress highlighting the success of researchers working on hybridization between maize and *Tripsacum* (Éluard 1978). With this gesture—echoing Debord's readiness to revise his views and adapt to emergent opportunities—began two decades of labour to forge apomictic maize. The trend of opinion at the Congress, Éluard wrote, was towards exploiting the utility of apomixis for human food crops. Taking encouragement from the extensive but unsuccessful work conducted on apomictic maize in Novosibirsk—that he witnessed when Petrov invited the ORSTOM team to Siberia—the transfer of apomixis into maize now became his prime objective. The Frenchman placed his first proposal for an apomictic maize programme on the desk of the director-general of ORSTOM, Yvars Ballester, in July 1979, less than a year after the Moscow trip, with ambitious plans to conduct the research at CIMMYT. The Russians had failed so far, he argued, because they had only one *Tripsacum* specimen to work with. Then there was the added complication of the overbearing Siberian climate. Even so, they had realized encouraging results. With CIMMYT's outstanding resources and the extended Mexican growing season, the holy grail of a major apomictic food crop seemed within reach. And although ORSTOM possessed extensive holdings of pearl millet, the actualization of apomictic maize would have far greater ramifications for agriculture, he proposed—a revelation in progress. It was the beginning of a long struggle to establish and execute an apomixis project. It was also a more frustrating 'dance of agency' than the *Panicum* programme ever involved, with unpredictable political economic and scientific challenges in play, and the resistance of unruly maize-*Tripsacum* plants to contend with, that in combination would eventually undermine any attempt at accommodation.

The proposal still sat on Yvars Ballester's desk in 1981 when Éluard presented his *Doctorat-ès-Sciences*, with a detailed study of Guinea grass. The director-general was on the examinations panel as was Dr Jack Harlan. This time, the maize plans were shelved—though rejection came with a counter-offer. For the next few years, Éluard could work on apomictic forage grasses in Brazil. But the geneticist and plant breeder was persistent. In July 1986, he wrote to CIMMYT directly with a revised proposal, but was turned down by the head of the maize programme. 'We have never worked on apomixis, we have no interest in apomixis, we will never work on apomixis', came the reply. But Emmanuel Marceau, a French colleague working on Brazilian coffee plants, was married to a Mexican national, and was interested in relocating to Mexico too. So the two men devised a project on *Tripsacum* diversity, whose hybridization with maize was of broad interest to plant breeders—with Éluard's apomixis research as a sideline. In those days, French government officials met annually to review overseas funding for CIMMYT and finally agreed to green-light the proposal in this underground format. The money was earmarked for the ORSTOM-CIMMYT *Tripsacum* Programme.

'So I had to hide my work on apomixis behind a nice picture that we are working on diversity of *Tripsacum* for maize breeding, something like that, you know?' Éluard recalled one Spring afternoon as we spoke in his office some 20 years later. We were in Marseille, among the umbrella pines and dusty industrial parks, at the futuristic building of Agromonde International. The French public sector organization for coordinating agricultural and scientific stakeholders sits on a winding street, across the road from IRD—to whose laboratories the ApoCORN relocated soon after Éluard's departure. ORSTOM and 'CIMMYT signed off on the *Tripsacum* programme back in 1987', he remembered. Back then, everything pure potential, and their plans for a future Apomixis Project, it was unthinkable it wouldn't come off. Éluard and Marceau's work commenced at CIMMYT's base outside Mexico City in 1989. 'Our research started in earnest just as the CIMMYT Applied Biotechnology Centre was completed'. But a new era was dawning, with genomic becomings and plant breeding deterritorializations on the horizon that would challenge the project's hypotheses—and new alliances of technoscientific practices and private sector capital to drive them. Such assemblages circled the emergent problematization that framed genomics as the most urgent and timely paradigm of the new technoscientific era. And this discourse would rapidly extend its disciplinary power and dominance.

Wide hybridization and the OCAPo

The ORSTOM-CIMMYT *Tripsacum* Programme resembles a Trojan Horse in hindsight, although complex research can often serve multiple ends. The five-year project approved in 1987 targeted the study of genetic resources and diversity in *Tripsacum*, and the potential for gene transfer to maize. While apomixis was a component in the proposal, it was not a primary focus. Initially, the programme proceeded according to the plan agreed with ORSTOM and CIMMYT. It commenced with a survey of Mexican *Tripsacum* populations, drawn from the established literature, Mexican *herbaria* and CIMMYT's collection of *Tripsacum* at its breeding station in Tlaltizapán, in the state of Morelos. This was supplemented by samples gathered during the project team's wide-ranging collection trips across Mexico. Most of *Tripsacum*'s known distribution area was covered, and a collection of 2,500 plants of Mexican origin established, with additions from geographical collections covering other regions of South America. The two researchers scrutinized the plants rigorously using the technology of the era. And there were intriguing results. Samples originating from Jalisco, near Tequila, and the area between Tuxtla Gutiérrez and San Cristóbal de las Casas, in Chiapas, proved to be particularly interesting from a scientific perspective.

Initial crosses between maize and *Tripsacum* commenced in September 1990 for training purposes. At that stage, the team consisted of Éluard and Marceau, with support from two PhD students funded by the French Ministry for Research, and several students from the Universidad de Chapingo and Colegio de Postgraduados de Montecillo in nearby Texcoco. But the lines between the *Tripsacum* programme and apomixis research soon became blurred. Éluard recalls having started crossing *Tripsacum* with maize experimentally as early as September 1991. And it quickly emerged that F1 hybrids were more suitable for hybridization with *Tripsacum* than pure lines. The team produced several thousand BC1 and BC2 plants over the next two years, using F1 hybrids as pistillate, and envisaged (confidentially) that a six-step backcrossing programme was needed to produce viable apomictic maize. This was important progress. Changes were also afoot at an institutional level within CIMMYT, reflecting wider developments in agricultural biotechnology, which played in their favour. Éluard went into print about his plans for apomixis in the 1993 CIMMYT *Annual Report*. 'The most difficult steps have been completed'—he reported—'we are confident of getting apomictic maize by 1997' (AR 1993). It was a timely call. With the appointment of David Penrose as director-general in 1994, CIMMYT entered an era where science and technology would take an increasingly high-profile role. On the one hand, this institutional change of direction emerged in response to new policy and funding vectors in the wider agricultural research and plant breeding fields (Murphy 2007). The *Tripsacum* programme, for example, now fell squarely in line with ORSTOM policy. And French officials responsible for approving project funding actively encouraged CGIAR to engage with the emerging new biotechnologies. This important policy shift among funders and donors would have major consequences for CIMMYT's activities in the years to come. But, importantly, the *Tripsacum* Programme was now supported by the new director-general. David Penrose was on a mission to co-opt the new biotechnologies for the resource poor and the public good (Penrose 1997).

By the 1994 review of the ORSTOM-CIMMYT *Tripsacum* Programme, the project team had successfully completed step three in their unofficial six-step programme to create apomictic maize. The project's external assessment was an important component of the review, and the report was overwhelmingly positive. 'At this point', wrote the distinguished American expert in plant breeding and genetics Richard Nefastis,

> the project looks like a space shuttle count down, proceeding on schedule. There may be delays, and difficulties, and they should

perhaps be expected, but the goals are very likely to be attained [...] If nature is willing, the additional work that has been proposed should succeed. (ER 1994)

While Nefastis' focus was on the project as a whole, the 'need to develop apomictic forms of economically important sexually reproducing plant species, including *Zea mays* [maize]' was highlighted as one of three key project goals. And with laudatory feedback, and enthusiastic support from other scientists and experts, the 1997 deadline seemed firmly within the project team's grasp.

Talking up the potential of apomixis for agriculture to entice potential funders was part of a range of seduction techniques that triggered an important expansion of the apomixis research field in those years, in terms of both capacity and ambition. Sympathetic expert reviewers were also a help, and, at times, a strategic tool in such efforts. The International Network for Apomixis Research, or APONET, the first international apomixis research network, had been launched, and would be the principal forum for leading researchers until the early 2000s. It capitalized on contacts developed from the Moscow Congress, and private sector experts joined the club to scout out promising research. Apomixis was no longer underground. In those days before the internet, APONET published the *Apomixis Newsletter*, which Éluard founded in 1989 and edited with US plant scientist and apomixis expert Charles F. Crane. The *Newsletter* carried scientific papers, contact details of members and subscribers, and details of recent publications, conferences, workshops and other news, with recipients based across the globe.[13] It also contained rousing and often polemical editorials, often by Éluard himself. And it was a highly effective platform for promoting the potential of apomixis for agriculture, that promissory rhetorical technique liberally deployed by apomixis researchers over the coming years. As a strategy, this was clearly a success, and confidence in the delivery of apomictic crops grew at an exponential rate.

It was an era when the leading projects in apomixis research were still embedded in plant breeding programmes and interspecific hybridization techniques, albeit in increasing alliance with the rapidly emerging biotechnologies. Wayne Hanna's team in Georgia was optimistic that it would soon have apomictic pearl millet (see Bashaw and Hanna 1990; Ozias-Akins et al. 1993). Another US Department of Agriculture (USDA) team under Bryan Kindiger was running its own maize-*Tripsacum* wide hybridization programme. Kindiger was working with Viktor Sokolov, Petrov's successor at Novosibirsk, who had taken advantage of the end of the Cold War and Yeltsin's chaotic presidency to reach out to Western scientists (see Kindiger et al. 1996). But the Americans also pulled no punches in talking up their work, which by now was squarely focused on the potential of apomixis for agriculture. So after positive reviews of the *Tripsacum* project and progress with interspecific hybridization, Éluard and Marceau published a flagship article. They argued that results from the *Tripsacum* Programme clearly demonstrated that apomixis would be a potent tool for maize improvement (Éluard and Marceau 1994). It was the first public call to establish a project to rival that of the Americans. Their proposal caught David Penrose's imagination, who was keen to advertise CIMMYT's technological credentials to funders (IR 1997:4, Penrose 1997). Importantly, for CIMMYT management, there was now a policy-driven justification for supporting apomixis research. Éluard and Marceau's timing was well considered—CIMMYT became a supporter.

So it was that Éluard's vision—inspired by the Russians, goaded by the Americans, both impeded and enabled by changing institutional agendas and wider research *dispositifs*—was finally realized. The OCAPo was established with a three-year mandate from 1994–1997, funded largely by ORSTOM with a contribution from CIMMYT.[14] Morale was high, the expertise of researchers on the team was increasing and support from CIMMYT management was robust. Securing research students, and early publications, had strengthened the project, and a Basque researcher, Gabriel Atxaga, would come on board as a fully-fledged member in 1997, funded by the prestigious Leverhulme Trust, with a project entitled 'Apomixis: The Small Farmer's Helper'. Alongside Mariana Rivera, a Mexican laboratory technician employed by CIMMYT since the early days of the *Tripsacum* project, this tight-knit team comprised the OCAPo nucleus until 2000.[15] As for the research agenda, ORSTOM-CIMMYT policy was to empower scientists to take the lead, and Éluard was appointed principal investigator (PI) in the new Apomixis Project. He determined how the research agenda would be operationalized and oversaw key decisions. (Emmanuel Marceau had left to focus on maize diversity and its symbiotic relationship to Mexican farming practices.) But this deferral to scientific leadership at the project level was complemented by a lack of scientific expertise on the part of CIMMYT and ORSTOM management. Scientists ran the research programme, but they did not call the shots when it came to the big financial decisions—a governance structure that would have important ramifications.

In the initial workplan, dating from January 1994, the project was officially labelled the second three-year phase of the ORSTOM-CIMMYT *Tripsacum* Programme. By March 1997, when the next review was due, it was known as the 'ORSTOM-CIMMYT Apomixis Project' (OCAPo). Its agreed goal was to introgress apomixis into maize by 'wide-crossing' *Zea mays* with *Tripsacum dactyloides* (gamagrass), its apomictic relative (Éluard and Marceau 1994). At the time, maize was believed to have originally been domesticated from the indigenous Mexican grass teosinte. So the logic of targeting interspecific hybridization with another Mexican grass relative is clear. Technologies such as flow cytometry, genomic in situ hybridization and large population screening were allied with wide-crossing in the quest for this first major apomictic crop for human consumption. An additional, technologically ambitious strategy, drawing on molecular genetics and championed by the team's younger researchers, aimed to isolate the apomixis genes in *Tripsacum* and transfer them directly to maize in the laboratory. In retrospect, it was a sign of times to come, but although 'candidate genes' were ultimately identified, molecular research remained on a modest scale and progress was limited (Valastro et al. 2001).

The emergence of the OCAPo research assemblage, then, relied on a complex series of translational processes, as different genetic and agro-technological knowledge practices developed within the CIMMYT institutional *dispositif*. From 1994 to 1997, the OCAPo targeted:

a) the production of the first comprehensive genetic analysis of apomixis;
b) the molecular mapping of hybrid progeny;
c) the study of novel maize-*Tripsacum* lines;
d) and finally, the transfer of apomixis itself, resulting in the planned commercial release of apomictic maize in 1997 (IR 1994: appendix iv).

The OCAPo was predicated on the conceptual proposition that apomixis was composite enough in genetic terms to be transferred within an interspecific hybridization experimental system. Yet the project could not have existed, for example, without the technological breakthroughs in plant tissue cultures during the 1970s that contributed to advances in modern wide-crossing technologies. Likewise, it was enabled by flow cytometry, a technological line of flight from that distant *dispositif* enabling investigations of human health disorders, which would now be used in rapid screening for apomictic maize progeny. The maize-*Tripsacum* model system acted as the project's foundational agent, whose cross-fertilizations via interspecific hybridization were, of course, inherently emergent manifestations of *l'intempestif* and the disjunctive force of nonknowledge (Bataille 2001). And the location of CIMMYT, and the cultivation station down south in Tlaltizapán, enabled and constrained project work through intensities of climate and infrastructure—while cultural factors, such as the close proximity of resource-poor farmers, fired the commitment to the open-source model of Apomictic Technology. Éluard's catalyzing role therefore echoes anthropological assessments of entrepreneurial agency in social transformation. It was facilitated by the research assemblage that he was instrumental in establishing. And it was shaped by the complex politics of plants resulting from interspecific hybridization method-ologies and everything *Gedankenlosigkeit* and otherwise that research practice involved. But the OCAPo was also embedded in transla-tional practices, of a globalized, historico-emergent character, which tempered and enabled its complex range of human-plant interactions (Ong and Collier 2005). Each territorialization and line of flight comprised pathways and obstacles, with disjunctive articulations, pre-cipitating a complex dance of agency and dialectic of resistance and accommodation in the OCAPo assemblage.

Meanwhile, external interest grew, as the genomics zeitgeist loomed on the horizon with the Human Genome Project as its figurehead. Enthusiasm for the new technologies spread across contiguous discip-lines and across the globe. For example, during the 1990s the Rockefeller Foundation, the Leverhulme Trust, the European Economic Community, the USDA, and the French and Mexican governments would all number among APONET's sponsors for conference and networking activity. Representatives were openly enthusiastic at the prospect of creating apo-mictic crops by the millennium. As the decade progressed, and the OCAPo ran smoothly, even Mexican politicians would visit. They were there to shake the hands of the apomixis researchers soon to work wonders with that jewel of Mexico's agricultural heritage, the humble maize plant, wor-shipped by the Mayans and many others, and a profound national symbol. If Tina Modotti had been alive, she would have photographed the apo-mictic corn when it finally came out, just as she did Mexican maize in her day, and Manuel Álvarez Bravo would no doubt have done the same. And it was as if the story of maize itself, domesticated by the peoples of meso-America some 10,000 years ago, was on the cusp of entering a second chapter. The prospect of its hybridization with *Tripsacum*, another indigenous Mexican plant, generated a low-level fervour among followers of events, as that new domestication event approached—harnessing the power of apomixis. In due course, it would also lead to modest funding from the Mexican state (see Figure 3.4).

Over the coming years, CIMMYT's management continued to use the OCAPo to showcase the institution's engagement with new technologies.[16] It remained a talking point for local politicians and, increasingly, GM activists. But an invisible worm was already at work. The decision to magnify the research in official press releases raised expectations concerning the timeline for delivery. Such anticipation would need to be reset beyond the encroaching temporal horizon, as the timeline extended, with a corresponding deflation of expectations. It is an intriguing echo of the deferred and contradictory temporal structure associated by anthropologists with 'everyday millenarianism'. Everyday millenarianism refers to religious and social movements that anticipate imminent change. Moments of rupture within such movements are continually deferred, and their expectations 'simply persist in being a crucial part of peoples' thinking throughout long stretches of time during which they have lived their millenarianism very much as an everyday affair' (Robbins 2001:527). Notably, participants go on with their daily lives while anticipating imminent change, in an apparent temporal bind. A similar sense of temporal life began to inhabit the Apomixis Project. In the case of the OCAPo, and other maize-*Tripsacum* programmes at the time, the 'new era' enabled by apomictic maize and all the changes to agricultural and economic practices that it entailed would loom up, only to retreat over the horizon.

This was one important temporal vector. But the timescape of frontier research within the OCAPo contained other potentially conflicting temporalities. Managing these conflicts presented challenges at times. For example, the repetitive time of raising thousands of hybrid maize-*Tripsacum* specimens was hard work. Cultivars needed to be fine-combed in a quest for the chance encounter with *l'intempestif* that could usher in the new world of apomictic maize—in a lab-based echo of Paul Debord's expeditions. In turn, the quest had to be recalibrated according to revised protocols resulting from the dance of agency, when success did not arrive. Such work called for patience, attention to detail and, ultimately, resilience. Meanwhile, the vortex time that converged dramatically at epochal moments, tied to project deadlines or rising expectations for the delivery of apomictic maize, generated instability and divergences in opinion.[17] That said, other aspects of the everyday routine of the project and lives of researchers continued, with their differing and interrelated temporalities—exemplifying enduring time just like one might expect. Managing these coexisting temporalities of practice within the OCAPo assemblage, and mediating their conflictive interaction, understandably generated pressure. In the medium term, slower-than-expected progress might also undermine the project's credibility, as it did within the apomixis field more broadly when funders took note that an apomictic crop still had not arrived. Reputations were also at risk. But these were not the only catalysts for the major changes to the project that were visible on the horizon.

The transversality of research heterocultures

From the 1990s onwards, other technoscientific assemblages began to colonize the apomixis research field, with a range of political economic vectors. One dominant *agencement* emerged from the intersection between corporate interests and the emergent genomics, driven by the desire to innovate characteristic of venture capital; the ability to bankroll expensive new technologies required for genomics biotechnologies; the desire for heavy profit; and the undermining of the public sector symptomatic of neoliberalism (Murphy 2007:137–140, cf. Sunder Rajan 2006). Such developments unfolded from the 1980s onwards hand in hand with the invention of advanced technologies such as PCR and automated DNA sequencing (Rabinow 1996). The impact of genomics on plant breeding advanced in the early 1990s, for example, driven by a significant expansion in research on epigenetics, which was of increasing value for endosperm production. These broader historical reterritorializations reached epochal culmination with the completion of the Human Genome Project and its fêting at the White House press conference of 26 June 2000. The accompanying 'postgenomics' zeitgeist that infused such developments, chiming with the febrile atmosphere that accompanied the millennium, was infused with overconfidence and hubris, prophesying an impending era of biotechnological achievements. The ideology was seductive, millenarian and vortex-like—but the reterritorialization of research was pure biocapital. In reality, such developments were grounded in the hard-nosed and business-minded genomics *dispositif* (Shreeve 2004). Perhaps the premature deadline for delivery of apomictic maize by OCAPo project scientists that identified 1997 as the first deadline for delivery reflected this wider optimism. Perhaps it also played to the audience of decision makers in ORSTOM and CIMMYT with a view to securing future funding. That said, the hegemonic fusion of classical and molecular genetics with plant breeding that dominated the quest for an Apomixis Technology in the mid-1990s was already outmoding in the eyes of the world of Ag-Biotech. Historical momentum was geared towards genomics. And revolutionary visions of apomictic 'crops that clone themselves' that appeared from time to time in the international press dovetailed with the utopian modernist vision of this brave new genomic world and its rupture with the limitations of the past, rather than the longue durée and 'enduring time' of plant breeding heritages (e.g. Pollack 2000).

The OCAPo was located at the juncture between CIMMYT's maize research programme and ORSTOM, funded research posts. Everyday control of the project rested largely with the PI, who also had a leading role in deciding the overall research agenda and striking agreements with the funders. Target end users were resource-poor farmers, in line with CIMMYT's mission and the project team's personal commitments, and reflected in the project's title: 'Equity in Access to Hybrid Vigor for Resource-Poor Farmers'.[18] The technology model was a 'natural' facultative apomict such as *Panicum* with 3 per cent sexuality. The goal was facultative apomictic maize with a residual capacity for sexual reproduction, enabling both cloning and selective breeding, which, theoretically, would maintain genetic diversity. A 'natural' outcome of introgression, the crop would avoid classification as 'genetically modified'. It would also comprise a de facto 'open-source' technology, in keeping with CGIAR's policy of producing 'public goods', as the goal was to release the product onto the market unpatented. The objective was to 'yield a final product that should be unconstrained by intellectual property rights, thereby guaranteeing its free access to developing world clients'.[19] Yet it was subversive in nature, as a facultative plant could be crossed with non-apomictic varieties of commercial maize, which could be rendered apomictic. As a result, an open-source cloning technology could be transferred to F1 hybrid maize, and in theory, generate deterritorializations and lines of flight in the world of the seed corporations. This would occur through a farmer selecting a sexually reproducing apomictic maize plant and crossing it with an F1 variety. The offspring of this cross would comprise a range of apomictic and non-apomictic maize plants, which could then be selected for apomictic plants with the desired qualities and backcrossed with the target maize. In this way resource-poor farmers could breed and fix cultivars for niche microclimates, and, in theory, farmers could clone F1 hybrid maize seed (Éluard, pers. comm.). Finally, the project's technical rationale was grounded in classical and, to a degree, molecular genetics, which suggested that apomixis had a relatively 'simple' genetic regulation controlled by one pair of alleles (Éluard 2000). Introgression into maize via interspecific hybridization was therefore theoretically feasible. In this way, scientific ontologies, plant breeding techniques and ideological agendas shaping product delivery were combined in a disjunctive alliance to produce a CIMMYT 'Green Revolution' innovation model for the biotechnological age—a research assemblage geared to cultivate vegetal becomings with a counterforce agenda.

'Our knowledge of apomixis in the wild was, OK, apomixis is facultative and it makes sense to keep it that way', Éluard would recall. We are in Marseille again. 'I mean, the private sector wants to have an obligate apomixis. They want to use apomixis in producing hybrid seeds [...] For the small farmer in the south, that's not the type of apomixis we want to give them'. The sky is blue outside his office windows, predictably. Such an idealistic crop would be difficult to harness for IPR. As a product of introgression, it would avoid tarring with the GM brush. Éluard nevertheless entered into lengthy exchanges with anti-GM activists during the 1990s, who viewed apomictic maize as both an ecological risk and a biotechnological tool for domination comparable to the infamous terminator technologies. Their concerns were probably well founded—with respect to some of the seed corporations then pursuing apomixis research, at least. General negative noise and rumour also increased concern among local farmers and activists in Mexico (see Éluard 1995, 2020).

As for the manner in which a facultative apomictic maize—the subversive kind—could be manipulated in breeding, or integrated with local seed systems, this was debated at CIMMYT and with other experts. One key interlocutor was Tom Hardy, a principal scientist on the CIMMYT economics programme, now a lead agricultural economist at the World Bank, who had published on seed systems and crop diversity. Debate was also informed ethnographically by Éluard and Marceau's field trips collecting *Tripsacum* samples across Mexico during the early 1990s, where they worked alongside resource-poor farmers and the indigenous peasantry. The plan for 'downstream implementation' (as it was termed) was to draw on participatory plant breeding techniques, and Hardy and Éluard drafted a report to these ends, although it was never published. They envisaged that subsistence farmers could cross local hybrids with apomictic maize cultivars to fix and retain heterosis—or they would introgress the apomictic trait strategically to conserve useful maize plants. In turn, the breeding tool would gradually be absorbed into local breeding practices and seed stocks via participatory selection and generalized differential transit, and, in this way, would circulate within informal seed systems. It was also predicted that the apomictic trait would spread from one farmer's crops to another's via open pollination. In this way, it might invade landraces and become part of the wider gene pool. This generated concern at CIMMYT—*inevitably*—although scientists played this scenario down, believing that 3 per cent sexuality would leave biodiversity unaffected.[20] Their confidence was chiefly based on previous research into apomicts such as *Panicum maximum*, and an unpublished commissioned study by a population biologist. As Éluard would recall would recall in conversation that blue-skied day:

> My objective is not to put 100% of apomixis in the field of the maize farmer, but to have a facultative apomictic maize which means that, for example, 80% or 90% of the farmer's progeny will be apomictic and composed of the best-looking ears that he selected the year before. The way I envisage that working is more like using the natural processes of selection as a model, rather than trying to force people to change and adapt.

In other words, plans for delivery and use by resource-poor farmers were tentative and would be firmed up once the product was in hand—this was *l'intempestif* they were trading with, after all, with uncertainties and emergent characteristics that were recognized as such.

Ultimately, the potential of a facultative tool also lay in the nomadic lines of flight that apomictic maize might generate within the wider agro-industrial and global landscapes of farming. These could benefit the resource-poor farmer, although unwelcome outcomes and reterritorializations were possible here as well. In this sense, Haudricourt's (1962) dualist thesis on the determining role of plant reproduction within immanent (Eastern-rhizomatic) and transcendental (Western-filiative) agricultural formations comes back into play. The apomictic assemblage of maize and *Tripsacum dactyloides* (a rhizome, as it happens) was endowed with the power by both CIMMYT scientists and corporate managers at the time to undermine the biocapitalist seed economy through subverting the transcendental filiation of seeds and enabling cloning (immanence). This disjuncture in temporal flow—or sexual deregulation at the biological level—would effectively create an immanent temporal moment and epochal boundary that halts the rupture between generations on which the political economy of F1 hybrid seed relies. For F1 hybrids, filiation by meiosis enables *discontinuity* across the generations, as we know, and thereby endows the filiative production of hybrid seed by corporations with economic value. In practical terms, the farmer keeps going back to the producer—the transnational corporation—Syngenta, Limagrain, Pioneer Hi-Bred and so on. S/he cannot reproduce F1 seed on the farm just by saving it.

Through preventing the dissipation of F1 seed that occurs through filiation, by replicating maternal DNA apomictically (immanently), farmers could thereby 'stop time', fine-tune temporal disjuncture, disrupt relations of commodity ownership and save viable seed.

'Clone-Time', then, is a temporal modality that deterritorializes sexual reproduction in order to create immanence—'remaining within'—in virtual terms. The power to stop reproduction at will also takes on an explicit and materially immanent character, whereby the plant reproduces by remaining in itself, 'without mixing' (see Helmreich 2007:294; Landecker 2005:2).[21] The causal relation of a nomadic facultative tool with wider disruption of the agro-industrial system, however, is not so deterministic as Haudricourt's original thesis or the conceptions of Éluard and other plant scientists might suppose. Or Deleuze and Guattari (1988), so too, with their neat dyad of 'royal' and 'nomadic' or 'minor' sciences—the first propping up capitalism and the state in its quest for reterritorialization, and the second subverting such hierarchies through cultivating becomings (see also Pickering 2010). Biocapitalism and its political economic siblings are experts at reimposing the molar identity of transcendence and process—holding onto the money—as anthropologists and political radicals know too well. And they are precise when it comes to reterritorialization of their interests.

Meanwhile, research at OCAPo continued throughout the late 1990s, a revelation in progress that was nevertheless on course for heavy seas. Charles Spillane (Spillane et al. 2001:1482–1483) provides a retrospective update from the front:

> Large population screening, flow cytometry, and genomic *in situ* hybridization were used to obtain BC_3 plants that have 20 chromosomes from maize and 18 from *Tripsacum*. Because little sexuality was present in BC_3 plants, the acquisition of subsequent BC populations was difficult. A plant with a chromosome responsible for apomixis has yet to be found among the BC_4 generation.

Scientific investigation at CIMMYT suggested that the maize genome significantly if not fatally altered the expression of apomixis among hybrids of the backcrossed generations. There were hypotheses for why this was the case, but little clarity. As Spillane (ibid.) continues:

> Specific attributes necessary for the proper expression of apomixis in maize could be related to gene dosage effects of specific modifiers or to parent-of-origin–dependent expression (i.e. genomic imprinting) of key regulatory genes that control embryo sac development and/or early seed formation.

In other words, work continued, methodically and professionally. But the plants were not compliant.

If the project's technical rationale was therefore grounded in classical, and, to a degree, molecular genetics, the political economic rationale for its facultative solution derived from the conjunction between CIMMYT policy, the globalized arena of agricultural biotechnology R&D and the team's leftist vision with a nod to the scientific utopian zeitgeist. From an anthropological perspective, what is notable about the research assemblage is its open-ended engagement with the field of nonknowledge generated by the technique of interspecific hybridization, and the dance of agency that accompanied this. Alongside this, while deadlines and conditions were imposed by funders and CIMMYT management, and the team itself organized its research into discrete periods, the articulation of activities was also driven by unruly plant agency and becoming. The project timescape comprised a number of articulated segments, which were flexibly coordinated and not directed by a centralized management infrastructure of committee oversight. Instead, under the PI's direction, there was responsiveness within the team to the outcomes of interspecific alliance, plant becoming and the vagaries of environmental influences, along with the recursive decision-making of researchers as they sought out *l'intempestif*. Likewise, the heterogeneity of the research assemblage was reflected in its transversal alliance of expertise and knowledge practices. Classical and quantitative genetics, molecular genetics, emergent genomics, wide-crossing and other plant breeding techniques, blended with Éluard's *Panicum* expertise, to generate an openness to emergent lines of actualization and flight. In sum, there was intensive articulation of different perspectives—characteristic of the 'effervescence' (Rabinow 2003:56) of this flexible *agencement* composite that gave a structural quality to heterogeneous relations where becoming and creative disorder were accommodated and cultivated (Marcus and Saka 2006:103; Rabinow 2003:56).

Anthropological insights emerge through consideration of how scientists and research assistants conceptualized maize plants, which reflected their respective alignment with genetics (in its various enduring forms), or the emerging genomics with the tremulous equivocation of its epigenetic accompaniment. If one believes that there is no *nature*—everything is conception—this takes on an intriguing guise as the only conclusion one can draw is that the scientists were talking about entirely different plants (Viveiros de Castro 2015:55–74). Despite these metaphysical fault lines, they evidently referred to maize plants using a uniform linguistic shorthand—*maïs* in French, *maíz* in Spanish, *Zea mays* in scientific shorthand and, in all likelihood, a range of situational appellations—in research practice and publications. This linguistic univocity both coordinated and concealed those radically differing ontological principles that resulted from their differential scientific paradigms and outlooks, and their actualization. Alongside the differing conceptions of scientists and research assistants, which presented under a naturalist bent, field workers (in particular) may have introduced non-naturalist indigenous Mexican perspectives. From a perspectival viewpoint, this coexistence of differing ontological foundations argu-ably constituted a form of hybrid multinaturalism—beneath the maize nomenclature rode different virtualities concealed beneath apparent similarity, becoming-other at differing tempos, whose actions would regularly resist the drift of knowledge practices altogether. They would even generate conflict and becomings within scientific practices and within the team, a *dark precursor* whose obscure and seemingly random system of reproductive operations triggered disjunction and differen-tiation from afar.[22] These transversal relations acted both as both a generative motor for research and a disjunctive catalyst, which pro-pelled the OCAPo through real pastures of nonknowledge populated by thousands of unruly maize-*Tripsacum* hybrids generated by the highly efficient CIMMYT breeding apparatus (Éluard, pers.comm.; cf. Candea and Alcayna-Stevens 2012). If time was disciplined within the project, then, this took place in a flexible and tactical manner, with sig-nificant redistributions of agency, including at the level of plant time itself (Marder 2013). In sum, the OCAPo was an affective creative universe—probably. Its heterogeneity of effervescent conceptual worlds and emergent ethnographic practices therefore constitutes an important point of contrast with the processualist and disciplinary operation of knowledge practices within the densely sutured ApoCORN

Looking forward, in order to anticipate the transformations within the OCAPo that took place in the late 1990s, two developmental points of friction are notable. The first concerned the project's multidis-ciplinary team of scientists. By 1994, Éluard had been joined on the *Tripsacum* programme by two doctoral researchers, paid for by the French Ministry for Research, as we know. Another would arrive in 1997, funded by the Leverhulme Trust. The first two doctoral researchers were molecular geneticists, although one had acquired some postdoc-toral plant breeding experience. The third researcher was a lab-based molecular biologist. All three, as younger colleagues, were oriented towards the emerging molecular technoscience, which they aimed to import into the CIMMYT research context (see Carrère and Pavese 2001; Valastro et al. 2001). There were good professional and scientific reasons for such ambitions. They were also, arguably, the outcome of affective *Gedankenlosigkeit*, where enthusiasm and pragmatic absorp-tion with the development of careers has a significant role to play in guiding knowledge practices, and in the generation and maintenance of ignorance concerning sideshadows. And the early career researchers made the successful case for incorporating an OCAPo subproject to identify candidate genes for gene transfer, which created the platform for future engagement with genomics and everything to come. Éluard, by contrast, with his training in classical genetics and plant breeding at ORSTOM, and apprenticeship to Debord, was interested and open-minded. But he was less conversant with new molecular techniques and favoured interspecific hybridization (Éluard 2000).

The second point of friction derived from the project's institutional foundations. During the 1990s, management was keen to talk up CIMMYT's engagement with hard science, partly to attract donors and funding. The CGIAR was animated by similar policies. This became considerably important as the attraction of genomics augmented throughout the decade—with its associated expenses. The combination of plant breeding and genetics that marked the early years of the OCAPo had seemed technologically advanced in 1994, but the postgenomics epoch was encroaching rapidly. When ORSTOM was restructured as IRD in 1998, engagement with technoscience was even more strongly emphasized in the organization's mission statement (IRD 1998). Plant breeding began to look retrograde to IRD managers dazzled by the new genomics—that savage sideshow—with its utopian promises, who were anxious at the competition for resources posed by INRA and other centres of research. In this increasingly bullish landscape, there was still a brief moment when public sector stakeholders of all paradigms came together—molecular, plant breeding and so on. From 27 April to 1 May 1998, apomixis researchers united at the Rockefeller Foundation's Bellagio Center in Northern Italy and issued the Bellagio Apomixis Declaration. Signatories expressed concern that the 'current trend towards consolidation of plant biotechnology ownership in a few hands may severely restrict access to affordable apomixis technology'. They recommended 'widespread adoption of the principle of broad and equitable access to plant biotechnologies, especially apomixis technology' and 'the development of novel approaches for technology generation, patenting, and licensing that can achieve this goal'.[23] Their worries were largely focused on the interests of resource-poor farmers, and the views of the OCAPo were prominent in discussions. Those ancient pre-millennial heterocultural days ... The unity was short-lived.

There was a sure-fire genomics revolution underway, whose knowledge practices and powerful globalized political economy was soon to undermine the OCAPo's dominant genetics discourse and its rationale for utilizing interspecific hybridization. The transversality of research heterocultures—regents for an entire morning—their days were numbered. There were imminent transformations in institutional settings and policy emerging at CIMMYT and ORSTOM that favoured the new biotechnologies. There were clear points of contact and disjunctive alliances between these globalized trends and the OCAPo's institutional nexus that would rock the project's foundations. There were intellectual tensions emerging within the OCAPo's scientific team. In retrospect, the OCAPo was poised for reterritorialization as a public-private partnership, and this catalysed—auspiciously—when one of its early career researchers made an important research advance. It is time to explore this transformation.

Notes

1 The CGIAR was founded in 1971. It unites and coordinates 15 leading not-for-profit scientific and agronomic research centres across the globe, with a combined workforce of approximately 8,000 scientists, researchers and additional staff. As with CIMMYT, it draws on a wide range of funding, including governments, philanthropic organizations, the UN and the World Bank. See: www.cgiar.org, accessed 28 August 2020. In addition to extensive research activity at its site in Mexico, CIMMYT runs projects in 50 countries and has approximately 1,500 staff. For information on its activities, funding and history, see: www.cimmyt.org/about/, accessed 28 August 2020.

2 See: www.cimmyt.org/news/bill-gates-highlights-impact-of-cimmyts-drought-tolerant-maize/, accessed 28 August 2020.

3 'The form of agriculture that Borlaug preaches may have prevented a billion deaths'.   See: www.theatlantic.com/magazine/archive/1997/01/forgotten-benefactor-of-humanity/306101/, accessed 28 August 2020.

4 ORSTOM, the Office de la recherche scientifique et technique d'outre-mer (Overseas Scientific and Technical Research Institute), was founded in 1953. Its primary objective was to translate agronomic knowledge into international development projects. From 1984, the acronym designated L'Institut français de recherche scientifique pour le développement en coopération (French Scientific Research Institute for Co-operative Development), and in 1999 ORSTOM became IRD, l'Institut de recherche pour le développement (Research Institute for Development). See: www.ird.fr, accessed 28 August 2020. From 1984, its mission focused on scientific research and collaboration with developing countries in which France has interests. IRD has incorporated a greater focus on technological innovation.

5 See Éluard and Marceau 1994 for a review of earlier work on maize-*Tripsacum* hybridization. CIMMYT also ran a maize-*Tripsacum* hybridization programme in the 1970s. As with Harlan's work, the intention was

to enrich the maize genome for plant breeding purposes, not to introgress apomixis (IR 1994:2).

6   I draw on correspondence and ethnographic interviews with Thomas Éluard conducted at Agromonde International in Marseille, Éluard 2020, and archival, bibliographical and internet sources, to construct this 'historiophilosophical' account—to recall Lundy's (2012) term.

7   Ploidy refers to the number of sets of chromosomes in a cell. Simply put, the term 'diploid' indicates two sets of homologous chromosomes, usually comprising one set from the female and one from the male parent.

8   Pickering (2017) later revised his formulation to acknowledge that non-human actors could be more 'cooperative' and engage with humans in complex ways—as plants certainly do—to complexify the concept of the dance beyond a focus on 'resistance', an insight endorsed in this narrative.

9   EMBRAPA is the Empresa Brasileira de Pesquisa Agropecuária, or Brazilian Agricultural Research Corporation, a national, state-owned enterprise.

10 Deleuze 2007:231. As Deleuze and Guattari write: 'becoming is not an evolution, at least not an evolution by descent and filiation. Becoming produces nothing by filiation [...] Becoming is always of a different order than filiation. It concerns alliance' (1988:238).

11 See Talcott 2019 on Georges Canguilhem's philosophy of error. In a suggestive passage for our narrative, Foucault (1998:476) writes: '[A]t the most basic level of life, the processes of coding and decoding give way to a chance occurrence that, before becoming a disease, a deficiency, or a monstrosity, is something like a disturbance in the informative system, something like a "mistake". In this sense, life—and this is its radical feature—is that which is capable of error [...] Further, it must be questioned in regard to that singular and hereditary error which explains the fact that, with man, life has led to a living being that is never completely in the right place, that is destined to "err" and to be "wrong".'

12 See Koltunow and Grossniklaus 2003 for a technical review that covers 'apospory'.

13 Three significant international meetings were also convened during the early 1990s—a USDA Apomixis Workshop in 1991; the first APONET Workshop in Montpellier, 22–24 April 1992; and the 1st International Apomixis Conference from 25–27 September 1994 at College Station, Texas, entitled: 'Harnessing Apomixis: A New Frontier in Plant Science'.

14 Detailed accounts and precise figures for the early years of the project are not available. Some financial data can be viewed in CIMMYT's financial reports (see https://repository.cimmyt.org/, accessed 28 August 2020). But this is incomplete and its interpretation is problematic. The level of funding, however, was substantially lower than indicated in Figure 3.3 for the ApoCORN.

15 Éluard and the French researchers were employed by ORSTOM; the Basque researcher was partly funded by the Leverhulme Trust. The project was also staffed by a number of Mexican postdoctoral interns during this time. In March 1997, for example, the team comprised Éluard, the three

European researchers, a biochemist from INIFAP in Mexico, a Mexican graduate student from Texcoco, six Mexican laboratory assistants employed by CIMMYT, and a field assistant at the Tlaltizapán station. Information on Mexican state and governmental engagement with the project is limited, but this clearly took place (e.g. see Figure 3.3).

16 For examples of this, and how expectations for delivery were raised by both team members and institutions, see CIMMYT AR 1993, 1996; IRD News Release 11, April 1996, 'Transmission of Apomixis to Maize by Hybridization: Researchers Nearer Their Goal'; CIMMYT News Release, May 1996, 'Scientists Announce a Breakthrough in Research on "Asexual" Maize', which was reprinted as the lead item in the CGIAR Newsletter 3(2). For a later example, see Charles 2003.

17 Morson (1994:163, 165; cf. Gurvitch 1964) writes of Dostoyevsky's scandalous scenes with their converging plots: 'Vortex time and sideshadowing work in opposite ways. If in sideshadowing apparently simple events [can] ramify into multiple futures, in vortex time an apparent diversity of causes all converge on a single [moment]. A hidden clock seems to synchronize this diversity so that, even though casual lines seem unrelated to one another, they not only lead to the same result but also do so at the same moment [...] The vortex strengthens as it approaches'. Apomixis researchers in the postgenomic era are more cautious about invoking the promissory vortex associated with delivery of an Apomixis Technology, preferring to deploy this more strategically to attract funders.

18 https://repository.cimmyt.org/, accessed 28 August 2020.

19 https://repository.cimmyt.org/, accessed 28 August 2020. See also Adams and Henson-Apollonio, 2006.

20 For example, Van Dijk and Van Damme (2000) offer a critical assessment of the dangers of an uncontrolled and uncontrollable apomictic trait.

21 Once again, this can be conceptualized in terms of a Spinozan principle of *causa immanens*. Or as Agamben writes: 'Immanence flows forth [...] [y]et this springing forth, far from leaving itself, remains incessantly and vertiginously within itself' (2000:226).

22 Deleuze (2004:119) writes, '[g]iven two [or more] heterogeneous series, two series of differences, the precursor plats the part of the differenciator of these differences [...] Thunderbolts explode between different intensities, but they are preceded by an invisible, imperceptible *dark precursor* ['précurseur sombre'] which determines their path in advance but in reverse, as though intagliated'.

23 http://billie.harvard.edu/apomixis, accessed 4 September 2010. The declaration was marked by its optimism for the impending delivery of an Apomixis Technology, and signed by leading researchers, many of whom remain active in the research field. It was published on the internet but is no longer accessible.

3 Disciplining time within ApoCORN

The private sector and the neoliberal bioeconomy

The OCAPo's development towards the ApoCORN public-private partnership was unexpected, but in retrospect, almost teleological, given developments in the late 1990s. A turning point emerged when one of the project's early career researchers made unexpectedly decisive findings. In February 1997, these formed the basis of the CGIAR's first patent application (Éluard et al. 1997), a major step for what was at the time an organization driven by core values of open access and public goods (which are strongly held by employees). The patent was ostensibly filed to safeguard the innovation for developing world markets (Éluard, pers.comm.). The filing outlined a method to identify a key nucleotide sequence involved in apomictic reproduction. It was in the nature of the frontier research OCAPo *agencement* that deterritorialization was inherent to its layout—that hieratic geometry—and emergent lines of flight could unravel the assemblage. They might even prompt reterritorialization within the orbit of magnetic systems—assemblages, for sure, but also the more powerful and enticing *dispositifs* of corporate biotechnology and, in particular, the new genomics. That kind of technology was beyond the OCAPo's financial resources, but would definitely enable the open-source mission. So a patent was not just a legal document staking a claim on food security in the future, or on sideshadows whose outlines could hardly be discerned—if indeed they would ever materialize—but a beacon flashing light. It said in the simplest terms that the promissory potential of knowledge within the OCAPo was now—or would very soon be—directly relevant to corporate scouts and all those hoping to realize the benefits of an Apomixis Technology (or blunt its deterritorializing edge). These scouts were often scientists themselves, with considerably more power than public sector PIs. Some no doubt aimed to attract researchers towards compliant technological modes that suited biocapital, and would engender its filiative reproduction.[1] Others were more open-minded and valued lines of flight, where these dovetailed with governing corporate interests. They were also, mostly, pathfinders for the emergent genomics assemblage that was, at the time, evolving as a fully-fledged *dispositif.* Which is not to rule out the fact that many of these individuals were committed to delivering global food security on their own terms. Éluard and others hoped, by contrast, that any alliance with biocapital—the seed corporations to them—would remain disjunctive and ultimately demonic (in our anthropological conceptualization) spurring maize-*Tripsacum* becomings that would develop sideways, rhizome-like, to invoke that subversive imagery. No equivocation here. And perhaps the signatories of the Bellagio Declaration aspired—*really*—to an imma-nent politics of plants, in Haudricourt's (1962) luminous terms, that would overdetermine the transcendent biocapitalist regime of global seed corporations in favour of the hardscrabble farmer and the hungry, once and for all.

Central to our narrative of this period of epochal rupture and reterritorialization within the OCAPo is a comparatively neglected phase of technology development. Social scientists have long recognized the need to analyze how technologies develop in the chaosmos of real life, with the hope of figuring out 'why did they actually take the form that they did?' (Bijker and Law 1992:3; cf. Deleuze 2004:57; Williams and Edge 1996). If such insights opened the door to studies of the 'social construction' of technology, researchers have tended to focus on proven technologies—or emerging technobait whose form is largely territorialized even if the research ends in eventual failure (e.g. Latour 1996).[2] I address here an earlier, protean phase in technology develop-ment, arrived at through conflict or failure during the provisional stages of a project—although Apomixis Technology is arguably frozen in an eternal return of provisionality. The conceptual term for this epoch is the *protophase*, when a technology's potential form is unclear, contested and the subject of speculation; research trajectories become deterritorialized and emergent; and the plane of nonknowledge is immanent and seem-ingly unbounded. The outcome may include a return to basic research, new lines of flight or even the abandonment of research altogether.

An archetypal example of a technology that remains in protophase is 'cold fusion', the attempt to initiate nuclear fusion (the sun's power source) without the extreme pressure or temperatures within our star. Such extremes prohibit solar fusion's use as an energy source—while research into its cooled counterpart has made limited progress, despite the lifelong efforts of some scientists. The protophase is an early stage of deterritorialization and retooling of research strategies that takes contingent ethnographic forms, where incidents of reterritorialization are short-lived, and relations within research assemblages are unstable (Parr 2010:70; Deleuze and Guattari 1988:508–510). It is a state of nonknowledge, for certain—but with attitude, even direction, oriented towards a hoped-for technological outcome—although it can still exist in a state of stasis, or confusion. In the case of PPPs, where actors with divergent interests are disjunctively combined, it can also involve significant political economic struggle over the forms a technology might take, particularly when these have heavy commercial ramifications. Social researchers and scientists alike agree that research trajectories adopted at early stages in technology development often have a lasting, or irreversible impact on the technology produced, and comprise a site of important strategizing (Bijker and Law 1992; MacKenzie and Wajcman 1999). Errors in decision-making at this stage—which may not become apparent until later—can also take on career-changing implications, for better or worse.

Ethnographic critique and conceptualization of the protophase in Apomixis Technology development is significant for two reasons.[3] First, there are different models for an Apomixis Technology, with distinct economic repercussions for stakeholders. Such models often develop along differing research trajectories, once initial, irreversible decisions are taken. Agenda-setting during protophases therefore has the potential to be politicized (Richards 2004). Secondly, Apomixis Technology development has oscillated between emerging phases and protophases on a number of occasions. The most significant returns to protophase took place during 2000–2004, as genomics displaced the late twentieth-century alliance of classical and molecular genetics with plant breeding paradigms, which dominated the field, to establish a new *dispositif*. This 'molecular turn' corresponds to increasing private sector involvement and the integration of apomixis research within the globalized bioeconomy. It produced a shift in the timeline for technology development, from short-term expectations to estimates of 20 years or more, as the challenge of realizing an exclusively GM technology hit home. And it heralded a change in technology end goals from open-source to patented technologies. This rupture was also symptomatic of the reterritorialization of apomixis research within that historical orbit in agronomy and plant breeding focused on the quest to control seeds and their genomes as commodity forms (Kloppenburg 2004; cf. Landecker 2007). More recently in the 2010s, after the ApoCORN's demise, a protophase of lesser significance emerged within apomixis research, with a low-level surge of public sector interest in hybridized genomics and plant breeding programmes giving rise to new lines of flight. Cue INRA, Dr Raphael Mercier ... something (potentially) tidal.

Also important, from this point on, both heuristically and in terms of their prominence within the ApoCORN, are the interrelated pathways of 'frontier research' and 'co-innovation'. Frontier research, as we know, is that term used by scientists and stakeholders for research focused on the quest for new knowledge, in which emergence and unpredictability are expected, and nonknowledge makes periodic raids and incursions into project work. Outcomes may be delayed or counterintuitive. Its dominant temporality is future oriented, and infused with the calculation and assessment of risk. Co-innovation refers to the 'cooperative' work of partners towards an agreed innovation objective, often stipulated in a legal contract. Where this involves collaboration between researchers and stakeholders with differing conceptual worlds, it also compels the actualization of transversal relations and novel alliances. In theory, both frontier research and co-innovation should intersect, and this is catered for in PPP contracts which, put simply, focus on the timeline of deliverables and specify collaboration in both services and financial terms—although the unpredictability of the former can complicate the actualization of agreed co-innovation goals. (History with emergence, again.)

ApoCORN emerged from negotiations held during 1998–1999, and an earlier period of informal discussion between CIMMYT, Thomas Éluard and the private sector. As an opening statement one can frame its emergence as a shift from a classic not-for-profit 'Green Revolution' R&D model, focused on production of public goods, to a neoliberal platform, allied with the genomics *dispositif* and increasingly oriented to produce biocapital (Helmreich 2008; Sunder Rajan 2006). In terms of policy trends, it is an early instance of the CGIAR's engagement with the private sector, which was endorsed and approved in the organization's Third System-Wide Review of 1998 (Paul and Steinbrecher 2003:110–112).[4] But it is important to grasp that the transition was significantly more complex and equivocal than it appears, and the goals of the OCAPo significantly influenced the PPP's early agenda. The private sector had been intrigued by the goings-on at CIMMYT for some time, and Éluard's research was of interest long before the OCAPo's had formally begun. Throughout the 1990s, at apomixis conferences, corporate scouts were monitoring public sector research for breakthroughs, and corporations had already inaugurated their own secretive research projects whose details will never come to light (pers. comm.; CHR). Éluard, a leading scientist and skilful publicist for his work (like most prominent researchers), was of special interest. The first meeting between Éluard and scientists working for transnational seed corporations took place in 1988. The Frenchman was in the United States for a year, working with US researchers to prepare for the ORSTOM-CIMMYT *Tripsacum* Programme, when he took a call from a scientist working for the oil company Shell. Based in Italy, the company researcher flew out to Florida to meet him. It was followed soon after by the first formal discussion concerning an apomixis PPP, which also involved Shell, and took place not long afterwards in France. There would be many such encounters.

At the conference sessions on apomixis that took place during the 1990s, private sector scientists and representatives were also present. 'They were people involved in scouting out what was cooking', Éluard remembered, speaking in Marseille years later, 'you know, anything in the public sector research that could be of potential interest for them'. Éluard's contacts with Dr Eduardo Bismarck, the influential scientist with Pioneer Hi-Bred, also date from this time. The transnational seed corporations employed scientists in prominent managerial roles to keep abreast of the latest developments and grasp their significance in both scientific and commercial terms. For such ends, they had decision-making authority where necessary—strategically, it was the right move. In turn, these scientists were fully integrated into the leadership structure—such as Bismarck himself, who built a successful career as both a director of research at Pioneer Hi-Bred and as a research scientist, and would one day be a representative on the Oversight Committee of ApoCORN. Dr Bismarck had been following apomixis research since at least the early 1990s, and potentially beforehand. The technical competence and impressive range of experience of such private sector scientists, and their authority within the corporate decision-making structure, contrasted markedly with that of public sector colleagues, who were PIs of research projects but were rarely (if ever) decision makers in institutional or financial terms. With the exception of Mike Metzger, the incoming director of the CIMMYT Applied Biotechnology Centre, managers of that era reputedly had limited knowledge of apomixis science and research, and the scientists who did had virtually no managerial authority. This was said to be particularly the case at ORSTOM and its later incarnation as IRD. It was a fatal disjuncture that arguably undermined the public sector model and would have significant ramifications.

Yet it is erroneous to characterize the serendipitous meetings of those distant days—whether at conferences or elsewhere—as solely expressive of private sector designs. Some would argue they were nothing of the sort, while others might suggest that public sector scientists could be just as calculating when it came to funding high-risk frontier research. They needed to be—each project was a gamble, looked at in those terms. Seed corporations collected information on the activities of apomixis researchers throughout the 1990s—that is indisputable. But in the case of the OCAPo, Thomas Éluard was on the lookout for ways of financing capacity building, consolidating his budding team and securing the project's long-term sustainability (Éluard, pers. comm.). He was fully aware that his research was potentially of life-changing value to corporate punters. It might even change the world. Researchers like Éluard also knew they needed substantial funding to realize their goals—in the case of OCAPo, ORSTOM and CIMMYT had limited resources and could not guarantee anything. There were multiple financial and political pressures on frontier research within the CGIAR, even in an institution such as CIMMYT known for its pioneering achievements with maize. Where global hunger is concerned, humanitarian priorities are ever present, and in some cases, as bills increase and results are awaited, colleagues and management may even count the cost in lives. When the OCAPo team talked with private sector representatives, then, they did so with their own financial and technological agenda, of their own accord, and with an eye to the palpable benefits for the project's agenda in terms of capacity, resourcing and access to technology.

Such motivations clearly reflect Spielman et al.'s (2006:1–2; cf. Byerlee and Fischer 2002:934) contemporary assessment of the perceived value of PPPs within the CGIAR:

> [which] are often considered to improve the management of resources by capitalizing on economies of scale and scope in research, exploiting complementary resources and capacities across the public and private sectors, and reducing transaction costs in the exchange of knowledge and technology.

And in fact, after a short-lived period of optimism during the early 1990s when apomictic maize seemed finally within reach, progress towards the OCAPo's goals had slowed. It was becoming clear to everyone that original estimates for delivery were in doubt and actualization an Apomixis Technology would need a few more years. But who would pay for it? As the 1997 deadline for commercialization became unviable, the need to rethink the timeline for delivery of an apomictic tool rendered the prospect of private sector resourcing more attractive. There was a clear-cut scientific rationale for the delay. Éluard would eventually publish a detailed analysis of the interspecific hybridization programme carried out at CIMMYT, concluding that it encountered difficulties due to the original transfer scheme between maize and *Tripsacum* (Éluard 2001:159–163). The major problem, he concluded, was down to the choice of a *Tripsacum* donor plant with a high rate of facultativeness. In technical terms, this had a solid foundation—its facultative disposition facilitated backcrossing of F1 maize-*Tripsacum* hybrids with maize. But it also destabilized future BC generations where a high degree of apomixis and low facultativeness were necessary to impart stability to the hybrid progenies. This contradiction could only be resolved by finding a better balance between the facultative qualities needed for interspecific hybridization, and the consistency required for backcrossing and stabilizing apomictic cultivars, so avoiding the actualization of alternative, unwanted sideshadows. Such complications once again demonstrate the importance of the protophase, and the temporal duration of decisions that are taken at that moment when nonknowledge overwhelms a project's identity.

Further technical issues also arose concerning a chromosome imbalance in the BC4 generation, linked to a difference in chromosome number between the donor plant and maize (Éluard 2001:161). As with the facultativeness issue, it was hard to say that anyone could have predicted it. And from nonknowledge—if the dice rolled in your favour—emerges possibility, as in time this insight formed the basis for a novel line of flight. This was Éluard's (2001) proposal that a fresh set of widecrosses, with a lower rate of facultativeness in the original donor plant, would be likely to produce the desired results. It was a plan informed by detailed reflection and which he would one day propose to ApoCORN managers with sure-footed confidence and informed by his years of experience with *Panicum maximum*. During the late 1990s, however, this novel pathway was not yet clear. In the CIMMYT labs and fields, unruly maize-*Tripsacum* hybrids did not comply with the best scientific plans to actualize apomictic maize—there was no sight of fully-fledged apomictic corn or anything like it. Meanwhile, expectations of success within a short-term timescale, sometimes endorsed by the project team and publicized by CIMMYT communications staff, were undercut by setbacks. The net result was an aura of dashed expectations about the pioneering project (AR 1994; CGIAR 1996). A virtual future of utopian scientific achievements encapsulated by an Apomixis Technology had snared the temporal modalities of the present into an impossible comparison with its promissory magnitude. At times, CIMMYT and the OCAPo were on the defensive.

But other scientific findings were still encouraging, reviews were favourable, and the field of nonknowledge was proving fertile. As a result, ORSTOM-CIMMYT signed off on a further three years of funding from 1997, with comparable research trajectories and a commitment to apomixis this time (IR 1997:3–10). Both managers and scientists alike agreed that, on balance, there was still cause for optimism. One of the OCAPo's researchers also produced those significant patentable findings, which in February 1997 formed the basis of the first CGIAR patent application. It was emblematic of CGIAR's changing attitude to patents, and, in a broader sense, collaboration with corporations. The patent was one of the first to target apomixis— Kindiger and Sokolov, and Wayne Hanna's team both filed the previous Autumn. And it did not go unnoticed. For the private sector, control of IPR was finally within reach, it seemed, and its representatives wasted no time in responding. Years later, the patent application's technical content is yet to bear fruit—in the public domain, at least—though in other respects the results were immediate, as a transversal alliance began to form between the OCAPo assemblage and new corporate partners. In this respect, we can note that 'public' and 'private' stakeholders within the rapidly emerging PPP were not distinct entities, imbued with discrete positivist identities and essentialized characteristics, but would be relationally constituted, politically and economically fused, and transversally differentiated, not least via the legal terms of the PPP contract. The resulting composite would comprise a novel alliance—an assemblage that would soon be stabilized and reterritorialized within the wider networks of corporate capital. And as reterritorialization progressed, it would be quickly absorbed within the emergent genomics *dispositif* that would ultimately reconstitute the OCAPo's heterogeneous multinaturalism.

Becoming public-private

The first contact came from Yoyodyne, Inc., which had ties with the head of the Applied Biotechnology Centre, Mike Metzger. A delegation of seven people arrived several weeks after the patent filing with a cheque for a million dollars—several times the OCAPo's annual budget. There was a catch—the money came in exchange for exclusivity. By this stage, Éluard had already alerted a contact at the French-based corporation, Limagrain, that Yoyodyne were interested, and the Limagrain representative had spoken with contacts at Pioneer Hi-Bred, with whom he was in periodic contact about those aspects of apomixis research which the two corporations were willing to trade or share. They quickly assembled, and there was a buzz at CIMMYT as a delegation of two representatives—one from each company—arrived shortly after Yoyodyne's team left, empty-handed for now. Limagrain and Pioneer Hi-Bred presented a more flexible proposal, promising less money but more scientific support and services, including direct access to high-end technologies and agronomic and genetic databases. The two corporations also sought exclusivity. The offer was better received but still unattractive to an organization committed to IPGs, and nothing was agreed.

David Penrose, the DG of CIMMYT, sat down with Éluard and Mike Metzger in the days after the landmark meetings to review the offers in detail. There were CGIAR values to consider, as well as the distinguished ethos of CIMMYT itself, and no one, to be honest, was at ease trading agricultural futures with the corporate Goliath. It became quickly clear that exclusivity would never concord with the CIMMYT outlook, and would invite repercussions from across the CGIAR, so one offer was easily declined. With Penrose's consent, the Yoyodyne cash offer was rejected. The other proposal, however, from Limagrain and Pioneer Hi-Bred, was scientifically intriguing and attractive for that reason. Meanwhile, news travelled about the talks at El Batán, back and forth across the corporate world. And given the hype that surrounded an Apomixis Technology in those days, it did not take long for another offer to emerge. A third party, Novartis—soon to be part of Syngenta—gave notice to CIMMYT management that it would agree to a non-exclusive agreement. There was room for manoeuvre now for CIMMYT and the OCAPo team. Limagrain and Pioneer Hi-Bred, sensing a shift in the terrain, relaxed their demands around the key issue of exclusivity. Negotiations—concrete negotiations concerning the gritty details of who would provide what, when and how much of it—could now commence.

In discussions with the seed corporations, IPR was a key focus for debate. CIMMYT and ORSTOM's binding condition was an intellectual property agreement that gave the institutions control of any future technology for developing world markets. And from intensive discussions a new two-tier or 'segmentary' licence emerged: Apomixis Technology would be furnished free of charge to farmers whose annual income was less than US$5,000, which, they agreed, encompassed the world's resource poor. This was only a starting-point, however, and in years to come the prototype licence was refined into a sophisticated segmentary model as the PPP progressed. Indeed, such was the importance of developing new IPR structures for apomictic scenarios that PPP partners put aside significant in-house finance for intellectual property management throughout the life of ApoCORN (Crouch 2009:3–4), although details are not in the public domain. In such a way, one can note, the ApoCORN's confidentiality agreement marks the boundaries of this narrative's own ignorance and unthought (Foucault 1966).

As for the legal agreement between the different parties, frontier research trajectories, timetables for delivery, governance structures, contributions and shared benefits in co-innovation were negotiated, agreed and detailed in the ApoCORN a priori contract (Crouch, 2009; see also Figures 3.1 and 3.2). Unsurprisingly, the form of an Apomixis Technology was unspecified, as this was expected to be emergent. The OCAPo team was intent on an open-source model, but research trajectories and their outcomes became the strategic, determining factor in which technological sideshadow would be actualized and the boundaries of ignorance and nonknowledge would be laid out. Clearly, CIMMYT could not compete with corporate partners in terms of financial resources or laboratory facilities. But the CGIAR institution covered operational costs for in-kind activities such as use of its extensive seed stocks and expenses for pre-breeding—that crucial period of plant cultivation that entails the identification of desired traits, and their incorporation into a cultivar prior to its production for distribution, which was central to the PPP's success. This would take place in Mexico, where a large part of the fieldwork would also be based. CIMMYT also funded research scientist posts in Mexico, in partnership with IRD (as ORSTOM would soon be known). And the potential value of the team's patent application—a catalyst for the whole enterprise—was also acknowledged. For their part, the private sector provided cash support and contributed services and new technologies. These included genetic materials for maize transformation, research conducted in specialist corporate labs and the supply of important promoters (Crouch 2009: 2–3).[5] This disjunctive alliance was an innovative experiment, and the agreement was the subject of interest in the industry, although the fine detail of contractual arrangements remains unknown. The ApoCORN was the only CGIAR PPP to focus on frontier research, and one of only two out of seventy-five to be structured around co-innovation over the next decade (Spielman 2006:3). But Byerlee and Fischer observe, PPPs 'often require considerable nurturing due to differences in business cultures' (2002:938). Such business models are ethnographically tailored and value-driven, which suggests diverse possibilities for conflict in terms of contingent ideals, goals and ambitions. The entry of the OCAPo into a PPP therefore placed complex pressures on research, linked to the embeddedness of corporate partners in global flows of biocapital. This hints at the political economic scenarios for subsequent conflicts in time.

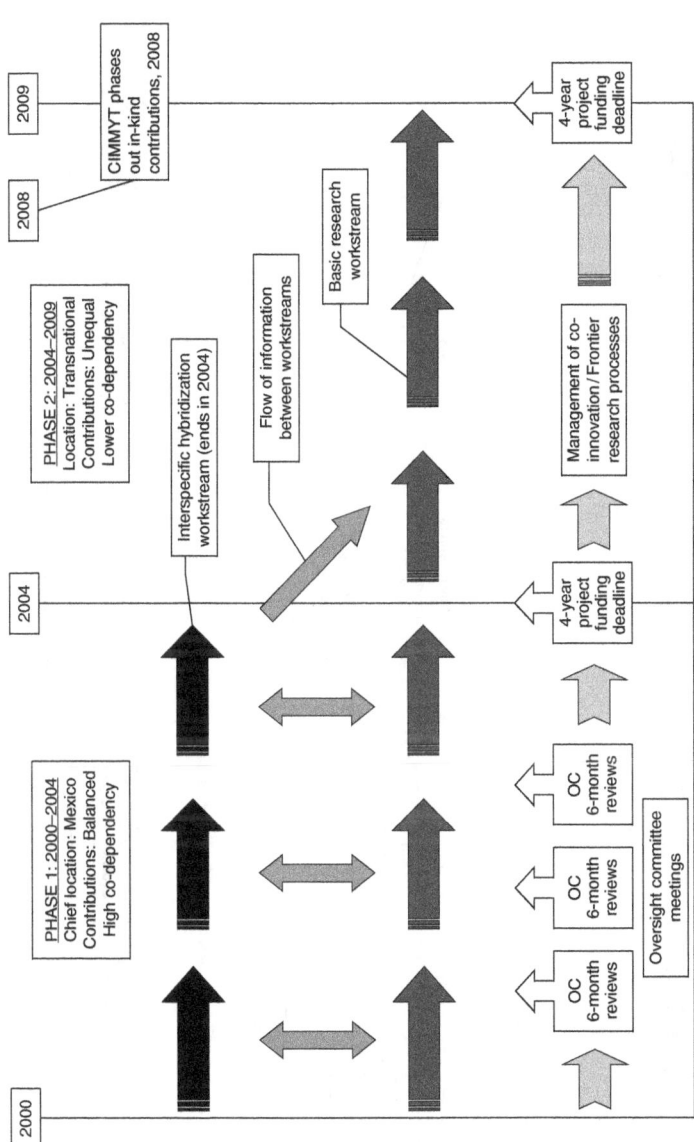

Figure 3.1 The timescape of frontier research and co-innovation activities within ApoCORN, 2000–2009

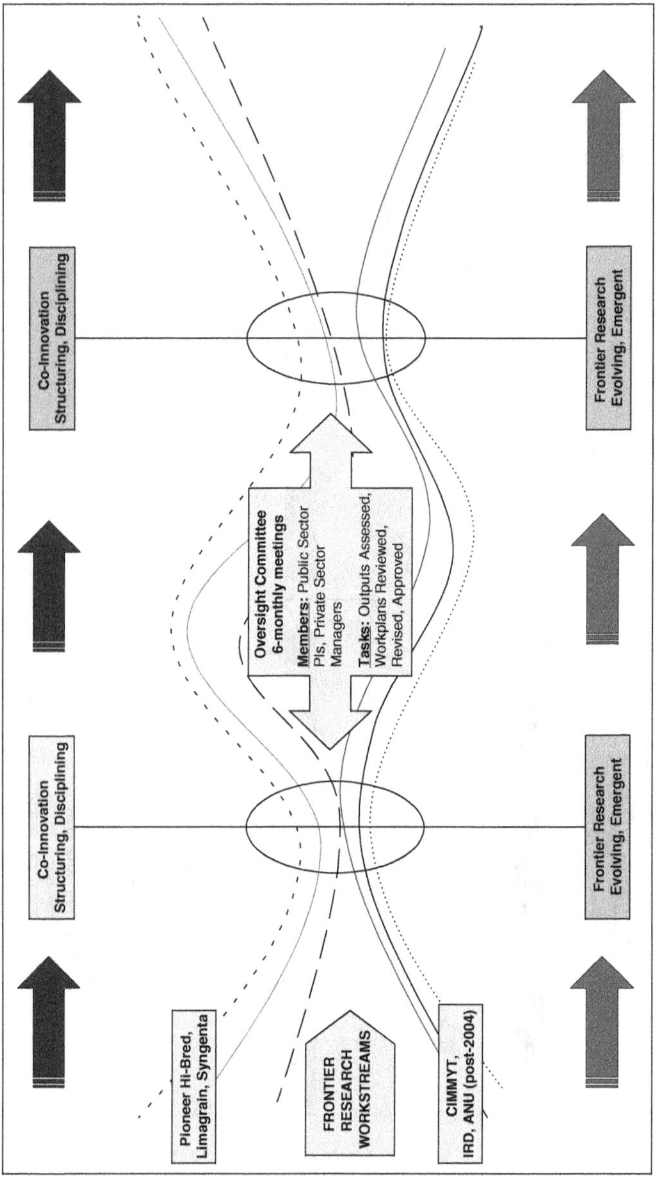

Figure 3.2 The disjunctive synthesis of frontier research and co-innovation within ApoCORN

Reflecting the joint-innovation model, for governance purposes an 'oversight committee' (OC) was proposed and set up when the PPP was inaugurated, which included Éluard, manager-scientists from each corporation, and the director of the CIMMYT Applied Biotechnology Centre, Mike Metzger. Apparently, the OC would aim to reach decisions by consensus (pers.comm.). Its task was to review progress and work plans on a semi-annual basis, ensuring feedback and a degree of flexibility in relation to frontier research goals and emergent opportunities. It also had a managerial role, with oversight for ensuring contractual agreements were adhered to. Eventually, the OC would also schedule meetings at each member's facilities in rotation, to take advantage of on-site expertise. However, it was two years before this complex agreement was in place, negotiated via a series of meetings at CIMMYT between private sector manager-scientists and lawyers in discussion with Éluard, Metzger and IRD and CIMMYT legal representatives. The extended negotiations also generated tension, which was particularly the case when it came to the legal dimension of discussions. IRD sent a lawyer to assist with no scientific background, who did not speak English, the language in which negotiations were conducted. Éluard spent the meetings translating the scientific discussions and explaining them to her, which was a challenging task. These complications generated tension with senior management in France.

In the first PPP contract, research goals were on an enhanced scale and did not significantly differ from the OCAPo, although the project's research capacity was greater and increasingly organized via processual and disciplinary idioms and structures. This mirrored a transition from the flexible organization of the OCAPo assemblage towards a coordinated biocapitalist *dispositif*. Research workstreams targeted the creation of apomixis in maize through the diversification of existing hybrid lines, in-depth research into endosperm development and gene-tagging.[6] However, a major obstacle that emerged from the OCAPo's final stages was the problematic formation of endosperm in apomictic maize-*Tripsacum* hybrids (Carrère et al. 2002). The underlying cause was probably epigenetic, and was of vital importance as the endosperm is the edible part of the maize plant. By 1999, the younger scientists were focused on understanding this imbalance. But Éluard viewed their research as merely treating the symptom, not fixing the underlying cause of this problematic development. He had already taken a broader view, and ultimately considered this to be linked to the original choice of *Tripsacum* donor. The plant variety in his view was too unstable to sustain the project across the multiple backcrossings needed to refine and stabilize the apomictic corn, due to the specific rate of its facultativeness.

The contract signed and sealed, there was a fresh sense of expectation as the new workstreams finally got underway. The PPP commenced its first phase of research in 2000 with a sophisticated structure and high profile as the CGIAR's first partnership with the private sector. The emergent qualities of frontier research were already laying the foundations for deterritorialization, as endosperm issues continued, but for now such challenges were to be expected, and governance structures were in place to facilitate rapid reterritorialization within the terms of the PPP agreement. The immediate issue, therefore, concerned how to proceed with the research. There were differing viewpoints, and the complex multinaturalist underpinnings of discussions across paradigms and the perspectives of different parties were apparent. But how would these correlate with the PPP with its new scientific and IPR priorities? Éluard argued that the logical step in addressing endosperm deficiency was to test his theory and open a new line of interspecific crosses, with a more stable *Tripsacum* progenitor. This could be selected judiciously, with a lower percentage rate for facultativeness, as a basis for future backcrossing and stabilization of the apomictic trait in an open-source format. This novel line of actualization could take its place alongside the endosperm research conducted by younger team members, with corporate support. But the proposal was not endorsed by the new partners, and importantly, Éluard was no longer the sole PI. This difference in opinion therefore combined with a change in governance to produce a deadlock. And it led in due course to the first point of conflict within the PPP and the alliance's first 'protophase' from 2000 to 2001. But before narrating that transition, and a second protophase from 2003 to 2004—in which an open-source Apomixis Technology was dispelled to the sideshadows—it is necessary to consider from an anthropological perspective the timescape of ApoCORN. This will lay the foundations for discussion of the politics of time that marked the development of the PPP over coming years.

Disciplining plant biomatter and the genomics *dispositif*

There were a number of conflictive temporalities or lines of actualization within the timescape of ApoCORN during 2000–2004. These were increasingly coordinated as part of a disjunctive alliance or *dispositif* where processual idioms and templates for practice were central to disciplinary practices. Helmreich proposes a 'formula to describe the making of biology into capital: B–C–B′, where B stands for biomaterial, C for its fashioning into a commodity through laboratory and legal instruments, and B′ for the biocapital produced at the end of this process, with 'the value added through the instrumentalization of the initial biomaterial' (Helmreich 2008:472; cf. Sunder Rajan 2006). Adapting this model to discussion of ApoCORN, its lines of actualization can be conceptualized in terms of the following criteria:

a) the generative agency and socio-technical disciplining of plant biomatter (B);
b) laboratory and legal political economic instruments for biomatter's conversion into public goods and commodities (C);
c) the impact of wider capital and cultural flows, within which the value of resulting products would be territorialized (B').

The first ApoCORN contract did not fully implement a biocapitalist research *dispositif* and processualist mode of production, partly due to the resilience of the OCAPo's goals and the continuation of features of its *agencement*. But subsequent developments gave rise to a new bioeconomic structure suitable for the production of biocapital. I present key trajectories, including their temporal structures, before discussing their interaction and its implications. Discussion omits the involvement of the Australian National University, whose participation began in the later phase of the PPP from 2004 and for which detailed information was not available. (By this point, research had relocated and CIMMYT's role was more limited.) Confidentiality agreements also shaped the boundaries of anthropological knowledge, and important internal data for the following discussion was kindly provided by CIMMYT.

I begin with interspecific hybridization, that key OCAPo research practice, which was initially placed at the heart of ApoCORN R&D before its relegation to the status of a sideshadow. Interspecific hybridization, as noted above, is a plant breeding technique whereby temporal emergence is methodologically thematized, when the unruly and disjunctive alliance of distinct genomes during meiosis is technologically facilitated by researchers. Emergence is also characteristic of, if not thematized in frontier research—a high-risk enterprise where results can be unpredictable, and unexpected sideshadows actualized as researchers engage with *l'intempestif*. The temporality of R&D in frontier research therefore benefits from a flexible open-ended governance and funding structure to accommodate these inviolate and integrated vagaries of emergence as a general expectation. As the length of time required for success is uncertain, the technique can conflict with timelines for the *deliverables* demanded by most PPP contracts in line with their commercial rationale (see below). In the case of apomictic maize, the temporal modality of an actualized real-world product might also prove subversive in political economic terms (cf. Haudricourt 1962). An end product produced *solely* by wide crossing would be a new apomictic variety of maize—not tarred and feathered with the GM label. It could not be controlled by a genetic use restriction technology (GURT) or currently available tools.[7] Instead, it would evince durability at the level of the genome's expression, or what one can term an *enduring* temporality (Gurvitch 1964).[8]

This temporal quality is extended here to conceptualize the materiality of biomatter that exists in a stable state—in this case as a result of the replication or 'cloning' of maternal DNA through seeds. As we know, this durability would still be subject to differentiation at the level of actualization, in relation to the intensive circumstances and environmental conditions under which the plant developed from the virtual Idea of DNA. But the ratio of difference to repetition would be reduced in an apomitic F1 hybrid maize compared to sexual plants, enabling continuity in key traits. In sum, the wide-crossing technique involves a relational, self-conscious 'dance of agency' between human actors, technology and the generative potential of two plant species, in the quest to produce apomictic maize. Its ethos can be conceptualized as one of 'revealing' (*alētheia*) through *poiēsis*, rather than via 'enframing' (*Gestell*) (Heidegger 1993b:318–328). Enframing is characteristic of the instrumentality associated with the practice of commodification integral to the production of biocapital, whereas interspecific hybridization is temporally resistant to this practice via its 'poietic' foregrounding of emergence and the agency of biomatter.

Furthermore, the end product would be a plant variety and public good with a transferable cloning technology. One dimension of its subversive potential lies in this apomictic capability. To build on earlier discussion, if a series of apomictic hybrids were released as public goods, for example, farmers could recycle these indefinitely and free themselves from purchasing hybrid seed every year. A further dimension lies in the potential to *transfer* the cloning technology through cross-pollination. If the apomictic tool could be switched on and off (via T-GURT), or the host plant's seeds were engineered to be sterile (V-GURT), it could be tightly regulated by legal instruments and controlled by corporate manufacturers. But a transferable open-source cloning tool is nomadic. It would enable becomings at the level of plant biomatter, and by implication, generate lines of flight in the hierarchies of agricultural social formations. It is not a 'royal' tool—the product of a royal science, as a GM apomictic maize might be (see Pickering 2010; cf. Deleuze and Guattari 1988). It would not enable reterritorializations intrinsic to industrial agriculture as these overlap with *dispositifs* of corporate and state power. Under such arrangements where 'research is increasingly modelled on the interests of states or private institutions, [the] researcher can become a political agent for the powerful' (Engebrigtsen 2017:51), and the centre of gravity is the production of biocapital. By contrast, in the eyes of OCAPo researchers, apomictic maize contained the potential for deterritorialization of these structures. The apomictic trait could transfer to other maize varieties, given that the objective was to produce an apomictic maize with 3 per cent residual sexuality. Farmers could clone F1 hybrid maize seed by introducing the apomictic trait through interbreeding. The tool would be nomadic, then, despite its tendency to clone.

The enduring temporality of an open-source apomictic maize would therefore render it subversive in relation to current IPR arrangements. It would be difficult to control the uses to which this technology was put, its emergent movement between maize varieties and landraces, and the 'unauthorized' recycling of seeds it could enable, through contemporary IPR regulation. In this way, it might indeed destabilize the political economic status quo, although control over the production and distribution of non-apomictic maize seed might still be guaranteed via patenting and plant breeders' rights. The resistance of apomictic maize obtained by interspecific hybridization to conversion into biocapital therefore exists at the level of both R&D operations—in terms of its unpredictable, open-ended engagement with the field of nonknowledge and inclination to subvert the disciplinary movement of PPP organization and contracts—and the relationship of the nomadic apomictic end product to IPR. For such reasons, Éluard's open-source model came to be viewed by corporate partners in the PPP as problematic. As Charles (2003:41) succinctly comments:

> [T]he corporate friends of apomixis are also its worst enemies. The reason is simple: seed companies have a financial incentive to keep self-cloning corn out of farmers' hands because apomixis breaks a natural sort of copy protection [...] CIMMYT's corporate sponsors might see topnotch apomictic hybrids from CIMMYT as competition.

In sum, the open-source approach was a product of a subversive 'nomadic' science, in that the OCAPo *agencement* combined a heterogeneous multinaturalist alliance of genetic and plant breeding outlooks within a flexible research structure and a sympathetic institutional nexus (see Pickering 2010). That said, it is important to note that this nomadism could also have a negative impact on existing maize varieties and land races, biodiversity and the environment. With the advent of ApoCORN, and the influence of the genomics *dispositif*, this trajectory was set to change.

Let us begin by looking at technological changes, which introduced important new capabilities for advancing a disciplinary *dispositif*. Researchers on the OCAPo had utilized molecular genetics in association with flow cytometry, a technology introduced to accelerate screening of interspecific hybrids for apomictic traits. This enabled a degree of technical instrumentalization, although it played a subordinate role and supported OCAPo technological goals. Within ApoCORN, however, a range of genomic technologies were introduced, facilitated by corporate partners (for example, AFLP-PCR, RFLP analysis; see Carrère et al. 2009:594). When deployed in relation to interspecific hybridization, these technologies accelerated that technical practice and expanded its scope (CHR). They also enabled novel interventions in the enduring temporality of plant reproduction, in relation to endosperm development, for example, which might eventually have rendered apomictic and other reproductive processes functional to commodification (cf. Carrère et al. 2002). Exploring the hetero-temporalities of plant time and, in particular, commercial fruit growing, Marder (2013:101) comments:

> In light of recent advances in biotechnology, it is possible to accelerate and, whenever needed, to retard the fruit's ripening—with the help of a gas 1-Methylcyclopropene or by engineering certain genes, such as LOV1 that control flowering time—and thus to harmonize the time of plants with the timing of the agro-capitalist processes of production and distribution [...] [which] guides us to the conclusion that rather than externally impose themselves, agro-capitalist techniques internally supplant vegetal potentialities and twist them, so that they obey the demands of the economic production process.[9]

The objective, arguably, was subordination to disciplinary procedures (Foucault 1977). Not all plants, however, are compliant, as we know, and may instead subvert human intentionality. Some scientists and corporate representatives made arguments within the PPP for the greater instrumentality and efficiency of new biotechnologies. This was coupled with an argument advanced by early career researchers, and endorsed by corporate scientists, that the project would stand a better chance of success if the end goal was a GM apomict (see Valastro et al. 2001). But corporate partners believed that a GM route would also secure technical and IPR control over any apomictic maize that was produced (corporate spokesperson, pers.comm.). Arguments were likewise made that the impact of apomictic maize on biodiversity could be better directed via a GM tool controlled by GURT (see Richards 2004:276). In this way, the vision of a biocapitalist ('royal') apomictic tool, one can propose, gained ground within ApoCORN and was set to triumph over the OCAPo's nomadic model.

This 'molecular turn' also exhibits close parallels with the phenomenon of 'molecularization'. As Rose (2007:36) comments: 'molecularization strips tissues, proteins, molecules [...] of their specific affinities—to a disease, to an organ, to an individual—and enables them to be regarded, in many respects, as manipulable transferable elements or units which can be delocalized'. The disciplinary programme of molecularization also has a temporal modality. Landecker (2005:2, emphasis retained) writes:

> These powerful techniques themselves belong to a genre of experimentation directed at *making cells live differently in time*, in order to harness their productive or reproductive capacities [...] [L]ongstanding genres of intervention in cellular plasticity and temporality are now moving from the background into the foreground of biochemistry and molecular biology, disciplines previously focused on knowledge of gene sequences and molecules in a more disembodied, atemporal fashion.

As a result of molecularization, which is characterized by its capacity to *rupture* the processual temporalities of cellular plasticity, 'natural' processes within ApoCORN were disciplined and made functional ('enframed') by new technoscientific practices, with contingent processual aims which correlated with the production of biocapital. In Gurvitch's classification of social time, this 'techno-cellular' processual temporality of disembedding and reterritorialization with a view to future utility resembles 'time in advance of itself, where [...] the future becomes present. This time [is] predominant in competitive capitalism' (Gurvitch 1964:33). 'Natural processes' were thereby reassembled into instrumentalized processual constructions. The temporal fragmentation associated with such instrumentalization is central, as is an inherent future orientation conceptualized as the outcome of a transformatory process. But molecularization also involves an emphasis on *promise*, and in turn, is indexed to 'promissory capital'. This important political economic relation comprises 'capital raised for speculative ventures on the strength of promised future returns', and is a key driving force in biotechological modes of production (Franklin and Lock 2003:6–7; Sunder Rajan, 2006:107–137). In this regard, and importantly, the timescape of molecularization within ApoCORN incorporated 'promissory' time, articulated with promissory capital.

Distinctive temporalities of molecularization within the PPP, then, were clearly embedded in socio-material contexts which were polychronic—comprised of multiple lines of actualization—but scrutiny reveals their dominant temporal and processual modalities. In sum, within ApoCORN, scientists increasingly used new technological practices associated with the genomics *dispositif* in both interspecific hybridization and basic research. In turn, these techniques operated to an important degree on *temporal levels* to intervene in, functionalize and refashion biomatter in accordance with processualist idioms and structures of governance, and processual understandings of biomatter. These enable the mediation and coordination of distinct activities within an overall, managerial structure focused on the OC. For corporate partners, these biotechnologies also offered the reassuring promise of the actualization of a GM technological sideshadow enabling patenting and production of biocapital.

The processual dispositif *of governance within ApoCORN*

These technical pathways unfolded within the political economic structures of the PPP. Contractually, ApoCORN operated on a four-year schedule running from 2000 to 2004, with a renegotiation and renewal of the PPP contract for phase two from 2005 to 2009. Partners agreed the frontier research to be conducted, and deliverables and contributions at the beginning of each phase, and these formed the basis of the PPP's legal contract and the foundation for co-innovation. The precise timing and extent of contributions appears to have been deliberated as research developed, although their total value for each phase was predetermined (see Figure 3.3). The four-year cyclical temporality of the PPP contract was therefore a key legal infrastructure underwriting the political economic pulse of frontier research and co-innovation, and structured its conflictive temporalities.

In managerial terms, ApoCORN possessed a processual temporal framework specifically designed for short-term disciplining and modelling of its disjunctive research alliance (see Figures 3.1 and 3.2). The OC governed in line with the co-innovation model (Figure 3.4). Meeting every six months, it coordinated labours of co-innovation and monitored and assessed the results of frontier research, which were communicated to the OC members through written and oral reports. Contractual enforcement took place via collective assessment of progress at OC meetings,

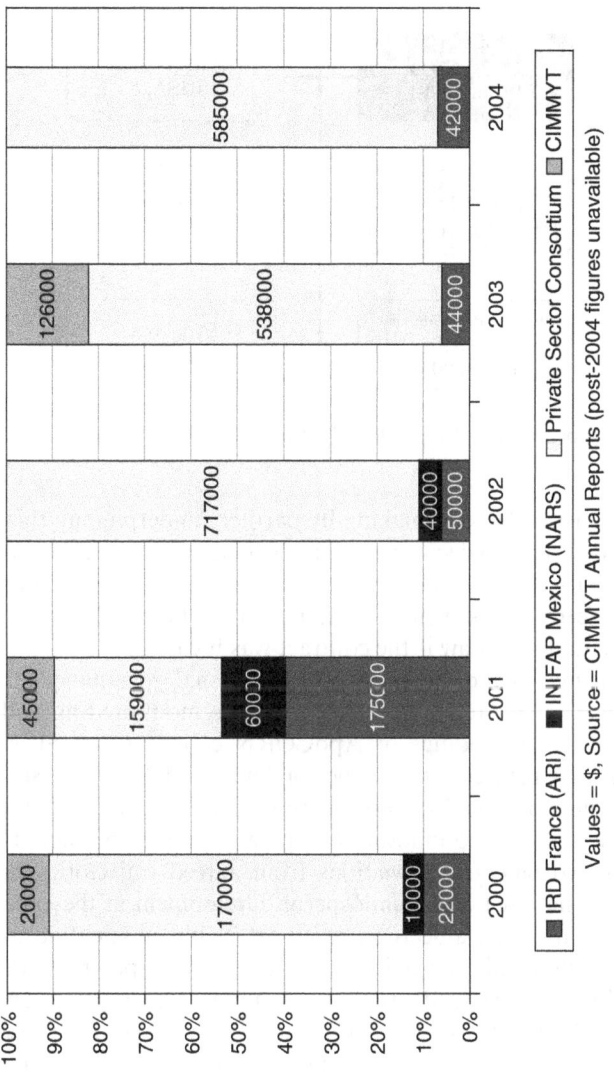

Figure 3.3 The allocation of member financing within ApoCORN during phase one (2000–2004)

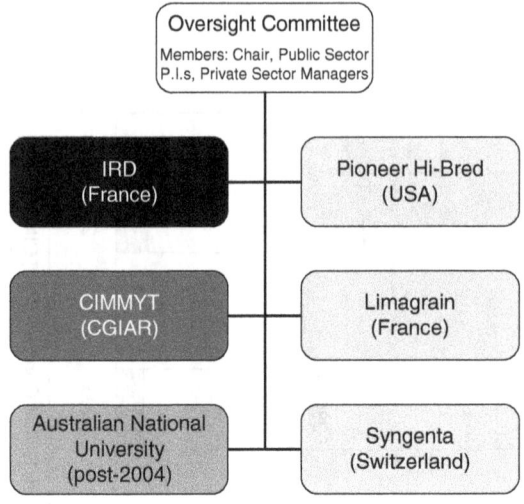

Figure 3.4 Governance of ApoCORN—the Oversight Committee

with consensual decision-making by partners underpinning this where possible, deferring to decision-making by majority if necessary. In turn, approved progress and delivery of contributions by researchers in relation to agreed targets ensured the payment of financial support—and partners could withdraw if the contract was breached.

To a degree, this disciplinary and processual governance structure enabled a partner's research trajectory to be monitored and adjusted in relation to the findings of ApoCORN colleagues based at other locations, and restricted the scope for lines of flight. Analysis at OC meetings therefore enabled co-ordination of multi-sited research in relation to the goals of co-innovation stipulated in the contract. The OC also endorsed emergent deviations from agreed trajectories, such as boosting basic research on endosperm development at the expense of co-innovation goals—a decision based ostensibly on scientific grounds, but with a political-economic rationale given the potential value of such findings for the production of biocapital. New research on endosperm development, for example, would be of value to corporations whether or not an Apomixis Technology emerged, and potentially to public sector partners as well. Ultimately, the OC advocated more significant changes, possibly towards dusk, such as eventual dissolution of the interspecific hybridization workstream. In sum, the OC imposed a disciplinary governance programme comprised of review, scientific debate and opinion-making, with a distinctive cyclical temporality that facilitated regular scientific and political economic intervention in ApoCORN's multifaceted research agenda. In this way, the OC could determine which research trajectories were actualized among virtual sideshadows and, importantly, which were not. Given the lack of specificity in the contract concerning the *form* of a potential technology, one can grasp how scientific debate in the OC therefore doubled as political contest over which technology remained *undeveloped*, alongside which Apomixis Technology might emerge.

However, OC meetings were also an opportunity for revelatory frontier research. They permitted combined expertise to be applied to scientific challenges. In this regard, meetings of the OC were eventually held at each partner's facilities in turn, to allow for consultation with on-site experts. For example, Edurado Bismarck was reputedly skilled at selecting company scientists to contribute to discussions (CHR). That said, other corporate partners were not so willing to share. Indeed, one informant close to the project claimed that, given corporate concern with confidentiality—integral to the protection of IPR and control of biocapital flows—the public sector was the only partner truly sharing information on a regular basis. The private sector was likewise perceived among CGIAR loyalists external to the project as contributing information of minimal value and using the PPP to harvest public sector data (Charles 2003:41; CHR). However, one public sector informant sympathetic to the private sector took the view that corporate partners were increasingly open as research continued (CHR). That said, only the corporations knew how much they really shared.

The OC *dispositif* permitted the close governance of frontier research and its integration with co-innovation within an agreed temporal framework. At this point, a comparison with the OCAPo assemblage is illuminating. OCAPo research only required systematic review by senior management at the end of each three-year funding period, with a view to securing continuation. There was no need to coordinate with partners or adhere to contractual timelines agreed with partners, nor were there legal limitations on how research could be reconfigured in response to unexpected findings. This was in keeping with long-term policy at ORSTOM—which funded most of the research posts—to defer to the judgment of PIs (pers.comm.). Likewise, there was limited potential for dispute over the political economic and commercial implications of distinct sideshadows, given the earnest commitment to delivering public goods. This enabled research to be conducted according to contingent and emergent priorities and opportunities, albeit within stalwart financial limits.

This was not the case with ApoCORN. The processualist temporal structure of its governance was organized around OC meetings, with a four-year deadline for delivery of contributions in accordance with binding legal contracts—although these were eventually renegotiated as frontier research suggested that initial targets would never be attained. A number of temporalities were actualized within the PPP timescape, with four dominant types:

(a) the emergent temporalities of frontier research (which in turn assimilated multiple cyclical and future-oriented temporalities structuring local lab and fieldwork activities);
(b) the cyclical temporality of OC managerial practices;
(c) the 'time pushing forward' of co-innovation deliverables and funding deadlines, which drove managerial imperatives; and
(d) the 'promissory time' of ApoCORN's co-innovation goal, apomictic maize.

Ultimately, interaction between these conflicting temporalities or lines of actualization was indexed to the distinct strategies and aspirations of partners, and surveilled through the disciplinary mechanisms of the PPP and its processual organizational structures. They were driven in turn by the desire to create an end product that could be mobilized as both public goods (for IRD-CIMMYT) and biocapital (private sector corporations). These were goals, of course, which did not necessarily overlap. We now turn to consider the wider flows and networks influencing such events, and their relationship to the genomics *dispositif*.

The promise and peril of an Apomixis Technology

In ApoCORN, specified research workstreams were the principal means by which the route towards a future technology took shape. Discrete research trajectories, enabled by different techniques, branched towards distinctive sideshadows: (i) an apomictic maize produced via interspecific hybridization, with subversive potential, and (ii) a GM apomictic maize likely to be controlled via a similar artifice to GURT. These models articulated with the different political economic configurations of PPP partners, and their interests in distinct global economic contexts and outcomes: (i) engagement with plant breeding scenarios in the developing world (IRD-CIMMYT), and (ii) globalized flows of biocapital (private sector). The deployment of an Apomixis

Technology in the future global bioeconomy therefore harboured distinct repercussions for different partners, depending on the form it took. On the one hand, these would emerge from developments within the confines and control of ApoCORN. On the other, they were linked to future markets and local regions of deployment over which partners had limited, or no agency.

The promissory temporality of R&D, in particular, intersected with interests of ApoCORN partners in a contingent, modal manner. As Fortun (2008:10) writes of biotechnological 'promise', ' "[p]romise" here [...] entails a mixture of a high degree of speculation, an avowed commitment stemming from multiple insecure extrapolations, and bets or gambles placed with a combination of care and risk'. Multiple contingencies in global markets and local contexts could affect future scenarios of deployment. A significant manner in which such contingencies translated into the strategizing of partners was in terms of 'manufactured risk'—of an ironic kind.[10] Corporations were not exposed to financial risk if the PPP *failed*. Their financial investment was low level compared to most research projects (corporate spokesperson, pers.comm.). Risk emerged from the possibility of *successful* co-innovation. Two virtual scenarios come to mind for corporate partners in disjunctive alliance: (i) forging an apomictic tool which might subsequently be used by another partner in commercial competition; (ii) successful emergence of products that could subvert market structures and impact negatively on biocapital—the nomadic Haudricourtian projectile (cf. Charles 2003:41). These risk scenarios were subject to cost-benefit analyses, with their economistic and processual temporal fabric, which in turn fired strategizing within the OC (CHR).[11] One can add to this the thought of what might emerge from the fields of nonknowledge and the unruly behaviour of plants, or the actions of *l'intempestif*.

ApoCORN kept scenario (i) in check via stipulations in the PPP contract, particularly confidentiality and proprietary agreements, which sought to curtail risk to biocapital of individual partners by ensuring that the 'trust' which enabled collaboration (alliance) was underpinned by legal sanctions. Recall Viveiros de Castro (2014:160): 'a becoming is a movement that deterritorializes the [...] terms of the relation it creates, by extracting them from the relations defining them in order to link them via a new "partial connection"'. Contracts curtailed the becoming-PPP that partners entered into. In this regard, there was open acknowledgement between partners that the PPP existed in a 'pre-competitive' stage, where co-innovation was possible given that an Apomixis Technology was still in development. Each partner was aware that this mode of cooperation might suddenly change if research was successful. Indeed, the contract specified what partners would acquire from ApoCORN research should a competitive phase initiate and the PPP be wound up (corporate spokesperson, pers.comm.). This pre-competitive temporality enabled the manufactured risk associated with future scenarios in which the project was successful to be temporarily neutralized, and co-innovation to proceed. The détente before the competition.

Scenario (ii) required pragmatic, ongoing regulation, as it concerned the potential consequences of existing frontier research trajectories, namely wide-crossing. This regulation depended upon strategic processual management of research, to guarantee that successful outcomes were favourable to members and, in the case of corporate partners, to hamper the actualization of undesirable technological sideshadows. In this regard, it was central to operational practice within the PPP. The key venue for managing this scenario of manufactured risk and its virtual potential impact on biocapital was the OC. As for *l'intempestif*, this infused such calculations and scenarios. Finally, such risk scenarios were less weighty for public sector partners, given their disengagement in many respects from aspirations to biocapital. Their ongoing interest in the PPP, however, was dependent on wider developments in the quest for an Apomixis Technology for the developing world. If research by the USDA into maize or apomictic pearl millet was successful during the lifetime of ApoCORN, for example, the IRD-CIMMYT contingent might have exited and purchased a licence on the new technology (CHR).

Other technoscientific practices, and the cultural ideologies and expectations associated with them, also influenced research. Of significance in the PPP's first phase was the growing reach of the genomics *dispositif* within the global biocapitalist mode of production. This was fuelled by the drive to innovate and seek out new knowledge practices and pathways to profit inherent in promissory capital; a related capacity to finance the expensive technologies enabling the genomics revolution; hopes and expectations of heavy profit; and the global retreat of the public sector with its heterogeneous effervescent research culture (Murphy 2007:137–140; cf. Rabinow 1996). Consequences for plant breeding began to be felt in the early 1990s, inspired by work at the Goldberg Lab at the University of California, Los Angeles, for example, where the processualism of epigenetics was configured as central to the study of endosperm development, and public-private collaboration precociously advanced. This historical flow (*dispositif*) realized its epochal moment with the sequencing of the human genome.

The genomics zeitgeist was characterized by hype and inflated confidence, prophesying a new era of biotechnological achievement to coincide with the millennium (Shreeve 2004). Such optimism, which was also prevalent among public and private sector apomixis researchers, buoyed expectations of delivery of apomictic maize within ApoCORN's first phase, and scientists also fed these expectations to senior managers to secure funding. But such sentiments played out negatively for advocates of interspecific hybridization. During the 1990s, CIMMYT management endorsed greater engagement with the biosciences to attract funding, reflecting CGIAR policy. When ORSTOM became IRD in 1998, science became more central (IRD 1998) and the synthesis of genetics and plant breeding that characterized the OCAPo quickly seemed out of date. Visions of 'crops that clone themselves' that surfaced in the international press from time to time (Pollack 2000) chimed with the utopian promise of genomics and its modernist rupture with the past, rather than the enduring time and longue durée of plant breeding heritages. These external temporalities of 'expectation' (Borup et al. 2006) reinforced arguments within ApoCORN that endorsed the processualist genomics *dispositif*. They stigmatized interspecific hybridization via its *allochronic* association with an 'earlier' temporal epoch—the pre-genomic era.[12] This lent weight to arguments for the need to invest in research with the 'best chance' of success within agreed timelines—namely genomics (CHR). Lines of flight leading to biocapitalist reterritorialization … far from the OCAPo quest and vision tinted with nomadic overtones. In sum, this illustrates how complex global pressures and expectations could translate into ApoCORN decision-making, via cost-benefit analysis with its focus on the time-value of funding (Nas 1996).

Conflicts in time within ApoCORN

The preceding historical ethnographic critique and anthropological reconceptualization presents the key conflictive lines of actualization which structured frontier research and co-innovation in ApoCORN. Let us now explore their interaction and disjunctive alliances, focusing on two key crises or *protophases*. The ApoCORN thereby shifted decisively to a partnership focused on the production of biocapital, and for corporate partners, the manufactured risk associated with the PPP's end goal was diminished. As Helmreich (2007:288) writes: 'Biocapital would emerge when laboratory instruments could be calibrated with market and legal instruments.' We now trace that pathway, with Osborne's (1995:200) observations as a guide:

> How do the practices in which we engage structure and produce, enable or distort, different senses of time and possibility? Whose futures do they ensure? These are the questions to which a politics of time would attend, interrogating temporal structures about the possibilities they encode or foreclose, in specific temporal modes.

The first protophase occurred during 2000–2001. It had quickly become apparent that problems remained with interspecific hybridization. There was acknowledgement within the OC. But there was no consensus on what to do. Éluard believed difficulties were caused by the OCAPo team's original selection of a highly facultative *Tripsacum* donor to initiate the hybridization process.[13] This was generating instability in subsequent backcrossed generations, and was, he argued, the cause of slow progress (Éluard 2001). Éluard proposed they select an alternative *Tripsacum* progenitor and start a new line of wide-crosses. This would deliver a more stable hybrid and potentially resolve problems with endosperm development. Within CIMMYT, however, other scientists disagreed, arguing that genomic technologies were more promising (cf. Charles 2003:35–36). It is also possible that the idea of another lengthy round of fieldwork was unattractive, and genomics viewed as important for engagement with the wider research field and the prospects of younger researchers (Éluard, pers.comm.). An impasse was reached, and the sense that the project was on the cusp of delivering a revolutionary technology gave way to concerns that it may have been misconceived. With no clear vision of the way forward, research assumed a protean form.[14]

Until now, ApoCORN remained in a first, transitional period of disjunctive alliance. New political economic structures were in place, promising articulation with the global bioeconomy. But R&D practices remained under the nomadic sway of the OCAPo. Éluard's team were focused on producing public goods, come what may. Biological materiality (maize and *Tripsacum*) had not been deterritorialized via new practices into ethical and ontological (molecularized) forms conducive to the production of biocapital (Helmreich 2008:465–466, 472; Rose 2007). The aspirations of corporate partners speculating with promissory capital on a potential Apomixis Technology were not being met. And the subversive sideshadow of Apomixis Technology development—an open-source apomictic maize—was still a goal. As a result, associated manufactured risk was not disciplined, processed, contained. The ideological differences at the heart of the PPP's research culture was now exposed (Charles 2003:41).

This deterritorialization event intensified in mid-2000 when Éluard mooted his strategy informally to OC members. There was a strong scientific rationale for the plans: interspecific hybridization is dependent on periodic 'tuning' due to its thematization of temporal emergence (Éluard 2001; cf. Pickering 1995:20). The USDA programme on pearl millet had explored several donor options since the 1970s, for example, and the PPP's timeline was modest in comparison (Bashaw and Hanna 1990). At the OC, however, genomics was viewed as central to progress with co-innovation. A GM maize in which apomictic reproduction could be switched on and off was desirable for its utility in breeding programmes (Charles 2003:41). This would potentially enable the legal use of GURT technologies, so that any potential impact on seed markets could be controlled. New findings on maize endosperm would also be a success in terms of frontier research, even if the co-innovation goal of apomictic maize was not achieved. Also significant was the fact that the unruly, emergent temporality of interspecific hybridization had already complicated the timeline for deliverables enshrined in the PPP contract. Plant-time was proving unruly.

Let us review ApoCORN's political economic timescape at this moment. (1) Temporal and processual structures of OC regulation of frontier research, enabled by six-monthly meetings, correlated with (2) suitability of the molecularization of plant biomatter for future IPR; (3) external pressure from the temporalities of expectation associated with the genomics *dispositif*; and (4) conflict between the temporalities of interspecific hybridization and the contractual timeline of the PPP, to generate a protophase marked by conflicts in time. This provided an opportunity for strategic improvisation with emergent research trajectories. The primary motivations for the strategy of corporate members were the production of biocapital, and management of manufactured risk. The outlooks of early career researchers were influenced by the belief that genomic technologies could deliver success (Charles 2003; Valastro et al. 2001; Éluard, pers.comm.). Éluard, as noted, favoured a new line of wide-crosses, with a view to production of open-source apomictic maize—that nomadic sideshadow—a solution to which all OCAPo members were arguably committed. Cost-benefit analysis regarding the time value of research was a key nexus articulating the disjunctive synthesis of co-innovation goals, contractual obligations, funding and partners' strategies, with temporal and political economic structures.

Further historical and emergent factors undermined the case for a new line of interspecific crosses. Éluard's decision-making authority as PI was now curtailed by the OC. Corporate representatives, by contrast, were scientists with managerial posts within their companies, and empowered with greater authority. There was no room for manoeuvre within the PPP contract to actualize emergent sideshadows such as Éluard had identified, nor sufficient managerial support to drive through an amendment. The proposal was out of sync with hegemonic PPP temporalities for delivery of co-innovation goals. It was in allochronic conflict with wider temporalities of expectation associated with genomics. Most importantly, perhaps, it was incapable of delivering the inherently temporal manipulation of cellular temporalities required for a patentable technology (cf. Landecker 2007:232–233). By the end of 2001, the scientist had left for a senior post at Agromonde International. As a result of improvisations with the temporal and political economic structures of ApoCORN timescape, emergent lines of flight and the complications of working with *Tripsacum dactyloides* ref. 65–1234 (if he was correct), the sideshadow was dispelled.

It was not yet the end of interspecific hybridization, which was the contractual focus of co-innovation in the PPP until 2004. Enough momentum remained in the workstream to reinvigorate it with new technologies. Reorientation comprised, one can propose, a negative instance of relative reterritorialization, where research plans were reformulated within original contractual guidelines with a view to continuing the workstream.[15] Research focused on applying genomic technologies to analyze deficiencies in the original line of wide-crosses. From Éluard's perspective, looking backward, the interspecific hybridization workstream was now 'pre-programmed' to fail. Indeed, given the problems that had been identified, one could argue that the sideshadow of an open-source technology had already been dispelled and manufactured risk contained (Éluard 2001). The ApoCORN's objective was now a GM technology, or nothing. CIMMYT documents present a more nuanced view (PW 2002). By 2002, the project was now coordinated by a research scientist who had joined during the early years, and named, within IRD-CIMMYT documents, Apomixis: Seed Security for Resource-Poor Farmers. Two main subprojects addressed the challenges that had emerged. The first focused on the development of apomictic maize germplasm. Negotiations had already begun with the Australian National University to develop activities related to the identification of genes of interest to artificially induce apomixis in crops. At this stage, outputs in the form of publications and conference papers were still progressing well. And more than 800 maize-*Tripsacum* lines had been produced and screened for apomixis during 2001, as the project continued to tap the creative forces of nonknowledge, although no apomictic plants were identified (PW 2002:F2/1).

Importantly, a second workstream focused on deployment of an Apomixis Technology, in a robust demonstration of practical commitment to resource-poor distribution. Research seemingly went much further than occurred under the OCAPo, with examinations of:

a) economic impacts globally and at the farmer level;
b) risks for biodiversity;
c) public and farmer acceptance; and
d) new procedures for breeding and seed multiplication (PW 2002:F2/1).

To achieve these concrete goals, the team sought out expertise from other CIMMYT projects, partners within ApoCORN, and networks were also developed through consultancies. Although there was not yet any prototype to trial, and would never be, information was gathered with a view to synthesis and further discussion, and other CIMMYT project outputs were also consulted. One example was drawn from the Oaxaca project, on geneflow and the dynamics of maize genetic diversity. The team also explored avenues for improving public knowledge on Apomixis Technology, including a feature with CNN Africa and contributions to several television programmes (ibid.). In terms of scientific advances, the emphasis had shifted, however, in a way that illustrates the differences in viewpoint on why the interspecific hybridization approach had not yet succeeded. For scientists remaining on the project, conclusions had already been drawn that, because of biological constraints—for example, at the level of genetic regulation, or seed development—the maize genome 'might not be an appropriate recipient for that type of apomixis' (PW 2002:F2/1). Research was still progressing—not through the development of a novel line of widecrosses. But the project was looking to another route, and 'assuming the risk that such an approach might not be successful, new activities that relate to the development of artificial apomixes have been initiated with the objective of producing maize varieties engineered for apomixis' (ibid.). Importantly, CIMMYT documents reveal, 'work plans established during [OC meetings] endorse such strategic issues'. Four subprojects were conducted across the ApoCORN supporting these enquiries. Two focused on 'diplospory', the natural apomictic type found in *Tripsacum*; one focused on artificial apomixes; and one focused on kernel development.[16] From the perspective of the IRD–CIMMYT team, all their decisions were led by best scientific judgement and practice, in line with IRD-CIMMYT priorities, and the results were clear. Interspecific hybridization was not delivering, and given the unpredictable nature of *l'intempestif*, the challenge was to decide where best and most productively to direct the project's resources. When I discussed this crucial transitional phase with the project's oversight director, Mike Metzger, many years later, at the CGIAR meeting at which some of this research was presented, this was also his assessment. Clearly, then, there were differences of scientific opinion with Thomas Éluard, and it was the plants who ultimately resisted direction and still retain the secret as to why they will not comply. But from an anthropological perspective, there are other lines of flight to trace.

A first OC review from 2002–2003 revealed that these obstacles persisted (CHR). The OC already supported basic genomic research into endosperm development and apomixis, but intensified interspecific hybridization research for another year. There was still positivity and goodwill to draw on from all partners. But success remained elusive, and the wide hybridization workstream was closed during a second protophase from 2003 to 2004 (CHR). 'Strategic' (basic) molecular research then became the PPP's focus, at which point, the Australian National University joined the partnership. Over the life of ApoCORN, publication of major outputs fell in quantity compared with the OCAPo, from greater than ten between 1995 and 1999, to approximately seven between 2000 and 2009. The resource-poor orientation in public representations of the project was gradually removed. And this ultimately led to CIMMYT's exclusion in practice, although not in name, during phase two of the PPP, as fieldwork was no longer the focus. CIMMYT management expressed gratitude, in fact, that its inclusion was maintained even when in-kind contributions ceased completely in 2008. Meanwhile, the boundaries of ignorance stayed in place around a nomadic open-source tool.

From this second protophase, therefore, emerged a final negative instance of relative reterritorialization and disciplinary transformation, where biopolitical transformation was achieved, and ApoCORN reoriented, largely towards basic genomics research on *Zea mays* and maize-*Tripsacum* hybrids. Recalling Helmreich's (2008:472) schematic model for conversion of biology into biocapital, let us summarize how this took place. Initially, the temporalities of biomatter (B) were modelled via interspecific hybridization (C), conflicting with aspirations embedded in co-innovation structures (C). This conflict restricted potential for conversion of biomatter into biocapital (B′). Reterritorialization via protophases, where temporal structures inherent in ApoCORN and external cultural and capital flows produced a focus for strategic realignment, enabled recalibration of technical pathways (C) to enable production of (B′). In this way, laboratory instruments were realigned with market and legal instruments (Helmreich 2007:288), and research reterritorialized within the longue durée of that biocapitalist political economy which strives to control seeds as commodity forms (Kloppenburg 2004). It should be added that biocapital (and apomictic maize) ultimately did not emerge, as biomatter proved resistant to recalibration—an enduring instance of unruly maize-*Tripsacum* time. In sum, a biocapitalist or 'royal' science operative within a processualist, increasingly monocultural genomics *dispositif* had replaced the nomadic, minor science and heterocultural *agencement* of the OCAPo with its combination of genetics and plant breeding.

To conclude, the motivation to produce biocapital was instrumental in the outcome of protophases, in particular the perceived value of a GM solution to corporations.[17] But it is controversial to argue that reterritorialization was solely the product of corporate intentionality. The project's rupture was complexly impelled. Indeed, in the final analysis, plant biomatter confounded efforts—wide-crossing was not successful within the allotted timeframe, and perhaps the OC invested in what seemed like the best route to success within the time available (CHR). If a solution was identified, a senior manager told me, it would definitely have been grasped, whatever the format (pers.comm.). It is difficult to conclude that this was not the case. In addition, later genomics research also cast some light on the reasons why wide-crossing was not delivered (see Carrère et al. 2009). Meanwhile, IRD's faith in genomics produced important results (Casci 2011), and its team, eventually released from the PPP, remained committed to a resource-poor solution (IRD 2011:1).

What is evident is how conflicts in time within ApoCORN's temporal nexus influenced the triumph of GM solutions, as opposed to lines of actualization that might have one day produced open-source technologies. In this regard, the evidence suggests that frontier research remains restricted when the private sector is involved, as corporations are focused on a temporality of short-term promises and financial returns that may be incompatible with time frames necessary for the understanding and restructuring of plant genomes. The desire of the private sector to manage manufactured risk with regard to future profits also necessarily delimits its engagement with the full range of sideshadows for Apomixis Technology development (cf. GRAIN 2001; Charles 2003). This has arguably influenced other apomixis research projects, where private sector corporations allegedly funded postdoctoral students in public sector laboratories with the intention of compelling these labs to sign confidentiality agreements that restrict information-sharing and collaboration with public sector partners. Such tactics, one reliable informant claimed, were agreed between high-ranking executives of different corporations to impede public sector research.[18]

This shift to a processual scientific idiom and practices of disciplinary deterritorialization and re-embedding was accompanied by a structural engineering of procedures, time frames and futures. It was mediated by legal contracts. The principal side-shadows of this monoculture were the alternative technologies and virtual futures that interspecific hybridization and the OCAPo heteroculture might have actualized. While processual idioms and templates were not absent from the OCAPo, they did not play a disciplining role, given greater flexibility in research practices and a related valorizing of emergence—arguably comprising an Arendtian 'ethic of the interval'.[19] By contrast, the procedures engineered by ApoCORN contracts and timescales ensured that the processual genomics *dispositif* remained dominant in a timescape comprising multiple trajectories, temporal modalities and tempos. Processualism, within ApoCORN, facilitated the disciplining of knowledge practices, enabling 'molecularization' whose goal was to render apomixis manipulable and create a biocapitalist tool. It was embedded in turn in a configuration of political economic relations—corporate dominance over the global seed industry. Here is arguably an example of how processual and organizational practices ally and interlink (see Arendt 1958:232–233). In this way, 'process' was discursively employed to weld a conflictive field of force and emergence into a disciplined transit through time towards a specific goal: a commodifiable Apomixis Technology, or nothing. A goal, one should add, that remains virtual due to the unruly actions of plants which, to date, have resisted such instrumental disciplining.

What is clear is the challenge facing CGIAR and other public sector organizations as they continue to consort with corporations, if emergent opportunities out of step and time with biocapitalism are to be seized for the public good. 'If you're not in control of everything, you're not in control of anything', Éluard commented on leaving ApoCORN (Charles 2003:41).[20] Writing of these biotechnologies, Rabinow (2008:4) observes:

> [T]here are other phenomena that are present today [...] that are emergent [...] [T]hat is to say, phenomena that can only be partially explained or comprehended by previous modes of analysis or

existing practices. Such phenomena, it follows, require a distinctive mode of approach, an array of appropriate concepts.

The ApoCORN was convened to invent; its goal was actualization of the virtual rather than realization of the possible. A politics of emergence was at stake, with actors strategizing under conflictive constraints to actualize favoured sideshadows from promissory futures. We have traced this trajectory from its precedents in the Soviet Union, through the heterocultural assemblage of the OCAPo—flexible, emergent—to the innovations of ApoCORN with its foundation in the monoculture of the genomics *dispositif*. Throughout, this narrative has examined the emergent *politics of time* within OCAPo and ApoCORN. And what of this conceptual focus? Given the hegemony of promissory time in ag-biotech research, perhaps this conceptual model is more broadly applicable. Translational medicine and research is one field that still seems affected by conflicting timescapes—coordinating the unruly temporalities of research on human biology or vaccines with the timelines of biocapitalist investment and expected returns is evidently challenging. Likewise, research on the promissory rhetoric surrounding the genomic *dispositif* might benefit from closer consideration of its differential temporalities. Increased attention to the politics of time— history with emergence, filiation with immanence—might complement Rabinow's 'distinctive mode of approach'.

Notes

1 In my case, I was approached in person by a director of R&D from a major seed corporation with the offer of access to PPPs with a major aircraft engine manufacturer (whose R&D director accompanied him) rather than conducting research on apomixis.

2 'On all these questions of innovations that never go straight, read the treatise that BL considers his best book on the invention and subsequent disinvention of an automatic Metro in Paris. The "love" of techniques is the practice of discernment'. Bruno Latour on Twitter, September 2020. See https://twitter.com/BrunoLatourAIME/status/1310121807660437504, accessed 28 September 2020.

3 I draw here on interviews with figures from the public sector and industry; analysis of scientific papers related to the PPP and apomixis research; articles in the scientific and international media; oral history transcripts; and ethnographic fieldwork. Corporate confidentiality agreements restricted access to certain information and informants. Given these agreements, information concerning the activities of the ApoCORN was restricted and only selected materials were publicly available. Some information was also

provided under the Chatham House Rule (CHR): 'When a meeting, or part thereof, is held under the Chatham House Rule, participants are free to use the information received, but neither the identity nor the affiliation of the speaker(s), nor that of any other participant, may be revealed' (www. chathamhouse.org/chatham-house-rule, accessed 28 August 2020). Where evidence is referenced to a personal communication, it should be noted that this is a reconstruction of events from the viewpoint of a particular individual and should be treated as such.

4 For a critical assessment, see www.grain.org/en/article/258-a-greener-than-green-revolution, accessed 28 August 2020. The CGIAR (including CIMMYT) has subsequently devoted much time and energy to considerations of how to make PPPs work for the benefit of the resource poor, despite its critics.

5 A promoter is a stretch of genetic material that acts as a switch for 'turning genes on'.

6 Medium Term Plans, CIMMYT, 2000–2002, 2001–2003 (https://repository.cimmyt.org/, accessed 28 August 2020). Gene-tagging is the process of identifying and tagging selected genes with molecular markers, thereby facilitating screening for them in cultivars and, potentially, cloning them.

7 In the case of 'V-GURT', use of GM plants is controlled by rendering second-generation seeds sterile; T-GURT permits seed saving, but genetic enhancements must be activated by a spray. Both are known as 'terminator' technology, and currently subject to a de facto UN moratorium and legal prohibition in Brazil and India.

8 For Gurvitch, 'enduring time' is where 'the past is projected in the present and in the future. This is the most continuous of the social times despite its retention of some proportion of the qualitative and the contingent penetrated with multiple meanings' (1964:31).

9 Helmreich puts it more bluntly: '[C]ontemporary biological science has become expert at stopping, starting, suspending and accelerating cellular processes, wedging these dynamics into processes that look like a molecular version of industrial agribusiness' (2007:294; cf. Landecker 2007).

10 For Giddens (2002:26), for example, manufactured risk is 'risk created by the very impact of our developing knowledge upon the world', that is, it is the product to a significant degree, but *not* exclusively, of human action.

11 By 'temporal fabric' I refer to cultural media used for evocation of temporal phenomena, and coordination of activities, for example, calendars, clocks or symbolic media such as language with its temporal markers (see also Gell, 1992:18–26). Data on the temporal fabric of the ApoCORN is restricted by confidentiality agreements—the boundaries of anthropological ignorance.

12 *Allochronism* is 'a systematic tendency to place the referent(s) of [a discourse] in a Time other than the present of the producer of […] discourse' (Fabian 1983:31). It can be used as a discursive strategy in establishing power hierarchies.

13 This refers to the 'donor' of the apomictic trait, that is, the plant which researchers selected to wide-cross with maize. The plant concerned—unruly—was *Tripsacum dactyloides* ref. 65–1234, from CIMMYT Plant Genetic Resources Center (Carrère et al. 2009:594).

14 This account is based on published sources, and interviews with Éluard. Other views from the OC were unavailable.

15 Deleuze and Guattari 1988:508–510. 'Deleuze and Guattari distinguish four types of deterritorialization along the twin axes of absolute and relative, positive and negative. Deterritorialization is relative in so far as it concerns only movements within the actual order of things [...] [and] negative when the deterritorialized element is immediately subjected to forms of reterritorialization which enclose or obstruct its line of flight' (Parr 2010:73).

16 Further scientific details of highlight results from 2001 can be found in PW 2002.

17 Anthony Cavalieri, vice president of trait and technology development at Pioneer Hi-Bred, put it bluntly: 'The thinking in the seed business is that apomixis would be more useful if you could turn it off' (Charles 2003:41).

18 Such rumours did not refer to ApoCORN and no inference should be drawn that any ApoCORN partners were implicated.

19 Arendt opposes an instrumental processualism to the disruptive character of human action, capable of initiating new processes, of which the emblematic symbol is birth ('natality'), suggesting that the value of processualism is ambivalent (Arendt 1958: 305–309; Braun 2007). Humans are not the only actors capable of intervening, as we know ... those unruly plants again.

20 Where the nomadic sideshadows of technology development threaten the interests of corporate partners, this may be true—although see Marianne Bänziger (n.d.) on CIMMYT's policy on PPPs, where that institution's commitment to delivering international public goods is clearly indicated. Bänziger was, until recently, CIMMYT's deputy director-general (Research and Partnerships).

4   Epilogue

Anthropology and the apomictic image of thought

This research has lain dormant for a number of years. Perhaps that was out of a concern for its findings. But there is no smoking gun. Cracking apomixis is a tough challenge. Plenty of public sector researchers are working hard on this, and more than a few in the private sector, and none has succeeded yet. There was suspicion about ApoCORN, acknowledged within the CGIAR. There was inevitably secrecy.[1] Behind these, however, there are trajectories and outcomes, pathways of actualization, lines of flight, even if these did not suit everyone's interests. There was *Gedankenlosigkeit*, on the part of plants. Perhaps on the part of some scientists. When isn't there?[2] There was competence, and dedication, and idealism, and hard work. There was disappointment. There was corporate politics. In 2010, I completed a review of a PPP. Some of that work was openly commissioned by the CGIAR as part of a review of its policy on collaboration with the private sector. This book is not 'about' that public-private partnership and its predecessor. Ethnographic and other kinds of information about it have been drawn on in the writing of this book, but details have been changed, and if these are now considered not as pseudonyms—which 'stand for' another, real identity, be that institution, corporation or individual—but experimentations with the anthropological imagination, which 'stand in', the distinction is clear. This narrative, rather, is an exercise in anthropological fiction, in Viveiros de Castro's (2014:187; cf. Strathern 1987) precise understanding of that term, or 'an experimentation with [anthropological] thought itself' (*Gedankenexperiment*). Different discursive practices—anthropological discourse, the OCAPo and ApoCORN—are brought into disjunctive alliance through the actions of a dark precursor (Deleuze 2004:119). The outcome for apomixis research is a becoming-anthropology—and for anthropology, the reverse. In this sense, real life is that dark precursor [*précurseur sombre*], the differenciator which lies beyond and propels this conjunction. The result moves beyond either pole. It resides where history and emergence meet, in a zone, one can propose, that comprises a discursive *equivalent* to the practices of interspecific hybridization. This book, then, transposes the events on which ethnographic research was based into a distinct medium, where difference is the operative word (Strathern 1988; Viveiros de Castro 2014). It constitutes an improvisation parallel to the historical—it is *un devenir*, but, importantly, with history as its reference point.

What pathways does this narrative trace, then? To summarize, it is a critical enquiry into the potential social forms of PPPs. It explores the potentialities of apomixis research. It expands upon the unwelcome consequences of working in collaboration with the private sector. It reviews the archetypes of organizational structures that are suited for frontier research. It dissects and renovates anthropological concepts of history and emergence—it explores the foundations of such concepts. It is not *about* real people or real research projects. It takes those as a starting point for a discursive anthropological enquiry that seeks to explore the concrete possibilities for frontier research, and the imaginative scope of plant scientists ... and how nonknowledge is engaged, by them. And how *l'intempestif* goes untamed. It is a cautionary tale. It is a tale of endurance—the endurance of research scientists. It is *not* an endorsement of this potentially fateful scientific research field. It is also about anthropology, and how anthropologists can reworld their ethnographic experience within discourse, for other ends—and the images of temporality that might assist in this. And for such ends, the apomict has some interesting lessons, as does its obviation of 'process' from the anthropological repertoire of foundational concepts.

History and emergence ... the apomictic image of thought. Throughout this book, the narrative has also circled a central paradox identified in understandings of history and emergence. Emergence, Deleuze proposes, or what he terms *le devenir*, 'becoming', 'is not history; history designates only the collection of conditions, as recent as they may be, that need to be overcome in order "to become", to create something new' (Rabinow 2011:62; cf. Deleuze 2007:231). The apomict, I have proposed, in its unique temporal configuration, enables us to mediate that paradox, which I have alluded to throughout the narrative, in a way that echoes other themes in this narrative. If we undertake a form of figural 'backcrossing' that sets the plant traits to one side to forge a 'pure' (anthropological) conceptual line, we can extract from this unusual and mysterious reproductive mechanism a novel anthrophilosophical (apomictic) image of thought that *re*produces while remaining within itself. History *with* emergence … transcendence *with* immanence … seemingly a contradiction in terms. And this novel image of thought signifies, perhaps, a conceptual means for imagining this equivalence, with the potential for further elaboration (cf. Viveiros de Castro 2014:77–93). It is a reworlding, which this narrative traces in its principles and in its form.

Let us advance these remarks, by extending our commentary on Hannah Arendt's critique of processualism, which is interwoven in a complex fashion with her guiding theory of 'natality'. Arendt viewed processual idioms and temporalities as operating on multiple levels in society—some positive, many negative. Processualism, she proposes, gained ground with the growing hegemony of Western scientific outlooks, but was simultaneously embedded in the expansion of capitalist economic organization, in which working activity is subordinated to end products and profit (Arendt 1958). It is also a key feature of so-called 'disciplinary societies' in the early twenty-first century (Hardt and Negri 2000). Processualism is also central to Arendt's concept of natality. 'For Arendt', writes Passerin d'Entrèves, 'the modern worldview is characterized by its emphasis on the idea of process, on the "how" of phenomena, be they natural or historical, and by the corresponding loss of the idea of Being' (1994:53). The natality concept, by contrast, highlights the human capacity to bring novelty into the world, thus disrupting the automatism of processes (and the temporal continuum) and initiating emergent acts and processes. The ontological fact of birth underwrites, for Arendt, this human freedom, and is echoed each time an individual introduces new action into the world. Natality is central to Arendt's critique of the hegemonic processual temporalities that, in her view, adversely fashioned twentieth-century cultural practice (Arendt 1958:97; Braun 2007:19–21). Ultimately, it enabled her to produce a philosophical outlook that displaced 'process' from its conceptual throne and to conjure a world in which alternative idioms and practices might crowd into view, and 'time' take a differential form. Arendt's approach is therefore multilayered, granting recognition of the value of the process concept—while enabling critique of its cultural hegemony.

An anthropology that is critical of processualism demands nuanced recognition of the temporally constructed nature of processes— and other genres of continuity, rupture and transformation, analytical and ethnographic—and an exploration of the consequences of such insights. It finds its foundation in a genealogy of thought that adheres to and informs the writings of philosophers such as Spinoza,

Deleuze, Foucault, Bergson or Nietzsche. The principal insight of such philosophers is an exclusion from conceptual schemes of any taken-for-granted assumptions of a transcendence of Being. All that exists of timespace resides and differentiates 'within' the living present. This might be conceptualized in terms of a Spinozan principle of 'imma-nent cause' which produces by remaining in itself, for example, or for some philosophers, as a 'plane of immanence' (Deleuze and Guattari 1994). And here the apomictic image of thought, as explored above in anthrobotanical terms, comes into focus.

As Agamben writes: 'Immanence flows forth [...] [y]et this springing forth, far from leaving itself, remains incessantly and vertiginously within itself' (2000:226). Nevertheless, as he points out, an aspiration to or invocation of transcendence cannot be wholly excluded from philo-sophical or social theories which adhere to the principles of immanence:

> [I]mmanence is not merely threatened by [the] illusion of tran-scendence, in which it is made to leave itself and to give birth to the transcendent. This illusion is, rather, something like a necessary illusion in Kant's sense, which immanence produces on its own and to which every philosopher falls prey even as he tries to adhere as closely as possible to the plane of immanence.
>
> (Agamben 2000:227)

In this regard, the illusion of transcendence is the corollary of any form of intelligible discourse, and Husserl (1966) provides one well-known, if problematic, model for how human consciousness subverts the imma-nence of being in time with his theory of 'internal time consciousness'. In temporal terms, then, a principal tenet of a temporally nuanced anthropology must be that discursive aspiration to temporal transcend-ence should be rendered self-aware, and subverted through acknow-ledgement of its immanent frame. History with emergence, the historian with the thinker ... The temporality of the apomictic image of thought, then, offers a conceptual route for mediating the tension that Agamben identifies—an apomictic image of thought.

I have illustrated how a major shift in trajectory within fron-tier research was enabled by a foregrounding of processual idioms in knowledge practices, and processual agreements for public-private partnerships. In this way the conflictive timescapes of research are mediated via legal agreement and a range of disciplinary procedures. By utilizing a differential temporal frame, it is possible to analyze how such processualism attains social form. In the case of ApoCORN, this revealed how processualism was a central component of disciplinary practices engineered to produce biocapital and exclude undesirable sideshadows. 'Processual practice' therefore operates to disembed and reincorporate intensities into pathways of actualization. We can recall that, when integrated with disciplinary programmes, this is a key dimension of practices of instrumentalization and rationalization, as exemplified in our discussion of ApoCORN. To be processed, to be disciplined, is to enter into procedure. All such disciplinary programmes arguably operate through making life available for re-embedding in processual cultural practices—that is to say, practices intended to create an ordered course of action or complex linkage between pasts and futures. Processualism, conceived polythetically, is perhaps the dominant temporality of the disciplinary society, in its many forms.[3] The outcome are projects, on numerous levels, for disciplining time.

Used as a doxic analytical frame, as we saw, process 'cools' the tensions of becoming. It obscures the virtual 'fullness of time' (Morson 1994) through its monological focus on constructing interconnections between successive actualizations 'over time'. It directs the unruly. By contrast, the task of anthropology is to render visible this act of mediation—while acknowledging that immanence, as 'reality in the making', must also elude anthropological practices of conceptualization and representation (de Beistegui 2010:192). The actualization of an event is immanent in time—time is not the transcendent measurement of the event. Time is likewise not a flowing or flux-like backdrop for anthropological analysis, but an emergent property of events. It is a differential multiplicity, materialistic, multi-vectorial, complex, aleatory. To create an interval in process, therefore, is an act of freedom in Arendt's sense—the time of the interval. Such intervals exist as perpetual side-shadows that many procedures may be said to work continuously to exclude.

In this respect, it is important to displace these fluid idioms—and employ shifting metaphors that reflect the topological qualities of timespace; or view the eternal renewal of metaphors as a method for combating the transcendental impulse and engaging with immanence itself (de Beistegui 2010; Deleuze and Guattari 1988).[4] In this sense, concepts no longer constitute empty forms awaiting content, or different representations of the same social reality, but are actively produced in analytical and ethnographic practice. Social life is no longer posited as existing within the 'flow of time', but as generated in an ethnographic zone that is focused on reworlding. Deleuze writes: '[W]e have no other continuities apart from those of our thousands of component habits', yet '[h]abit draws something new from repetition—namely difference' (2004:94–95). Yet we can extract the temporal modalities of apomictic reproduction to forge a concrete, apomictic image of thought from which to imagine such metaphysical assertions in terms of an equivalence. Conjuring such equivalent modalities of emergence and historical time, as we have seen, ultimately enables more effective anthropological purchase on conflictive, multiplex 'timespace in the making'—human and plant—within time's plasticity.

Frontier research and the politics of plants

The circumstances in which this anthropological research project ended its first phase may also raise eyebrows. This took place as the result of a review conducted by a former senior manager from one of the ApoCORN seed corporations, Mike Koteks, contracted as an external reviewer and chair of advisers—as Dr Nefastis was in his day. The ethnographic project was funded by the ESRC Genomics Network to explore how the social sciences could contribute to the deployment of an Apomixis Technology, and led by a co-signatory to the Bellagio Declaration. It was rapidly clear to us that there was no technology in the pipeline, but funding was in place and used to explore how the failure of scientists to make progress could become a focus. In turn, this created challenges for social scientific progress, as it became evident that talking about failure is not what scientists are disposed to do, to anthropologists or anyone else. One Spring day, in the attic rooms of the Georgian mansion which housed the ESRC Centre for Genomics in Society at the University of Exeter, a meeting reviewed this project's future pathways, chaired by Koteks. His recommendation was that there were better uses to be made of ESRC resources than to fund research on why apomixis research had not yet succeeded. It is an echo, perhaps, of that call made within ApoCORN regarding the end of interspecific hybridization. Several years down the line, it is difficult to disagree with this judgement, but the event itself illustrates the penetration of corporate actors within public sector research, and hints at the suggestive narratives their participation could generate and the probable need for firmer boundaries and greater independence. A number of individuals were nevertheless able to share their experiences, as we have seen, and with recent successes among heterocultural apomixis projects, these testimonies have turned illuminating with time.

In this light, the pioneering work sponsored by CIMMYT, and pursued by scientists from ORSTOM and IRD, serves as an example of how to take risks with *l'intempestif*, and develop flexible research assemblages that might one day deliver the new crops that apomixis researchers aspire to. That is not to say there are not better goals than self-cloning maize. Devotion of energies to organic and non-GM crops, as the OCAPo was originally designed to do, might be a wiser use of resources, for some commentators. But unless such questions are asked of apomixis in ways that might permit their development for common interests, rather than corporate ones, they will never be satisfactorily answered. The CGIAR has continued to work on models that will deliver such formats, as has CIMMYT, including cooperation with private sector partners, where goals and contributions are clear, and international public goods will be protected. While such collaborations are inevitable within the current global economic system, they are not the only model that needs to be developed if all lines of flight are to be cultivated. It is a lesson that the example of ApoCORN illustrates, along with how a small team of researchers can ask moonshot, utopian scientific questions, and with the right blend of enthusiasm, commitment and expertise, push these to the top of the agenda of institutions such as the CIMMYT, IRD and the corporate gene giants. The heterocultural assemblage of OCAPo also left scope for plants themselves to be involved and exert their agency. And as this agency is better understood, their collaboration might be better enlisted.

What form might this take? In a suggestive essay, Andy Pickering (2008) writes of the perils of assuming that we can resolve the flooding that periodically afflicts the Mississippi Delta, and can take on catastrophic proportions when hurricanes strike. The solution, the authorities propose, is to call in the US Military to strengthen the complex system of levees that shape that colossal river's course. 'Secretly, it was all being dictated by the needs of technology [...] by a conspiracy between human beings and techniques' (Pynchon 1973:529–530). A step change in our thinking, Pickering argues, would seek the solution in better accommodating human needs and desires around the river's resistant course—a more viable solution as the unruly twenty-first century develops. He sees the same 'minor' or 'nomadic' tendencies in the work of artists such as Willem de Kooning and their manner of working *with* the agency of paint. We also hear an echo of this conception of agency in the multinaturalism that Viveiros de Castro (2014:49–75) intuits in Amazonian peoples, similarly confronted with the vastness of the non-human and their place within it, where the other is actively pursuing its own ends—in this case, imagined in human terms—and these must be accommodated. The same, one could propose, applies to plants, which have a will of their own, and the outcome of the OCAPo–ApoCORN trajectory illustrates this. Working *with* plants and the environment, rather than assuming that we have dominion over them, is perhaps the most notable distinction in technical ethos and practice between the OCAPo and ApoCORN. We witness this in the 'nomadic' technology of interspecific hybridization and its thematization of temporal emergence. Increasingly, as we witness the ways in which Earth's systems push back against humanity and the anthropocentric outlook of the West, it is clear that this more cooperative way of working needs prioritization. A heterocultural approach, then, assumes we do not have all the answers, and must work with our limitations—the monoculture tends towards the reverse.

For the plants also generate a more than human politics, and need to be on board with the human project if solutions to global food security and the challenges of the climate emergency are to be found. It is that most unpredictable manifestation of *l'intempestif*, whose latest incarnation, in the form of the novel coronavirus, continues to ravage the globe—and accommodations must be forged if solutions are to be found. In that vein, let us end this book without closure, but leave the door ajar. And in a manifestation of that ethos of 'working with' rather than 'working on' that interspecific hybridization encapsulates—a counterpart to discursive equivalence ... history *with* emergence ... the historian *with* the thinker—let us ask the question. What if Thomas Éluard, with his proposition that a more stable *Tripsacum* apomict, allied with more advanced technological capability, holds the key to viable apomictic maize? This now seems unlikely. Many years later, he has arrived at a different conclusion and accepted that the interspecific hybridization of maize and *Tripsacum dactyloides* would not deliver an apomictic crop and the hoped-for, subversive Apomixis Technology (Éluard 2020). But a secondary theory he developed at the time, and discussed with a team of researchers at INRA in 1999, presents an alternative solution which he still endorses. This involved triggering the deregulation of sexual reproduction, through genetic transfer of the apomictic mechanism from *Panicum maximum* to maize. Progress at the time was blocked, it was rumoured, by government managers. But what if—in a counterfactual moment of reworlding—he was right?

Notes

1 Spielman and von Grebmer write (2004:27): 'What are the origins of this mistrust and suspicion? The confidentiality and nondisclosure agreements that accompany many public-private partnerships are a likely source of tension given how alien they are to public-sector researchers. These agreements not only prevent public researchers from sharing knowledge with colleagues but also generate suspicion among third party actors who observe or involve themselves in the public research agenda. For example, the secrecy surrounding CIMMYT's apomixis research [...] may contribute

to misperceptions and suspicions of the private sector and of public-private partnerships among some CGIAR researchers. [T]hese attitudes may also result from [...] an inward-looking, exclusionary attitude common among CGIAR centres and researchers.'

2 See Part 2 for further elaboration (cf. Bovensiepen 2020).

3 Hardt and Negri gloss: 'The disciplinary society is [...] constructed through a diffuse network of [...] apparatuses that produce and regulate customs, habits, and productive practices [...] [D]isciplinary institutions (the prison, the factory, the asylum, the hospital, the university, the school, and so forth) [...] structure the social terrain and present logics adequate to the "reason" of discipline. Disciplinary power rules in effect by structuring the parameters and limits of thought and practice' (2000:23).

4 As de Beistegui writes: '[I]mmanence always remains to be made, that is, conceptualized. This, however, does not amount to turning immanence into a concept [...] [T]he plane of immanence is never given as such, or fully intuited; it needs to be drawn through the creation of concepts. In a sense, such a task is never-ending' (2010:24).

References

Details of some items in this list of references are redacted as the names given in the book's narrative for their authors are pseudonyms. This indicates that the publication is real within the world of the book, but its identity is concealed, in line with anthropological best practice and in order to respect the privacy of the author/s.

Adam, Barbara 1998. *Timescapes of Modernity. The Environment and Invisible Hazards*. London and New York: Routledge. https://doi.org/10.4324/9780203981382

Adams, Stephen and Victoria Henson-Apollonio 2006. Defensive Publication: A Strategy for Maintaining Intellectual Property as Public Goods. The Hague: CGIAR International Service for Agricultural Research Briefing Paper, No. 53.

Agamben, Giorgio 2000. Absolute Immanence. In *Potentialities: Collected Essays in Philosophy*. Pp. 220–240. Stanford: Stanford University Press.

Altieri, Charles 1979. *Enlarging the Temple: New Directions in American Poetry during the 1960s*. Lewisburg: Bucknell University Press.

Arendt, Hannah 1958. *The Human Condition*. Chicago: University of Chicago Press. https://doi.org/10.7208/chicago/9780226586748.001.0001

Asker, Sven and Lenn Jerling 1992. *Apomixis in Plants*. Boca Raton: CRC Press. https://doi.org/10.1201/9781315137537

Bänziger, Marianne n.d. Interview on AAA drought-tolerant maize. Basel: Syngenta Foundation for Sustainable Agriculture. www.syngentafoundation.org/dr-marianne-banziger, accessed 28 August 2020.

Bashaw, E. C. and Wayne W. Hanna 1990. Apomictic Reproduction. In *Reproductive Versatility in the Grasses*. G. P. Chapman, ed. Pp. 100–130. Cambridge: Cambridge University Press.

Bataille, Georges 1985. Base Materialism. In *Visions of Excess*. Pp. 45–52. Minneapolis: University of Minnesota Press.

Bataille, Georges 2001. *The Unfinished System of Nonknowledge*. Stuart Kendall, ed. Minneapolis: University of Minnesota Press.

Bataille, Georges 2014. *Inner Experience*. Buffalo: SUNY Press.

Bear, Laura, ed. 2014. *Doubt, Conflict, Mediation: the Anthropology of Modern Time*. Oxford: Wiley-Blackwell.

Bennett, Jane 2010. *Vibrant Matter. A Political Ecology of Things*. Durham, NC: Duke University Press. https://doi.org/10.1215/9780822391623

Bicknell, Ross A. and Bicknell, Katie B. 1999. Who Will Benefit from Apomixis? *Biotechnology and Development Monitor* 37:17–20.

Bicknell, Ross A., Andrew Catanach, Melanie Hand and Anna Koltunow 2016. Seeds of Doubt: Mendel's Choice of *Hieracium* to Study Inheritance, a Case of Right Plant, Wrong Trait. *Theoretical Applied Genetics* 129:2253–2266. https://doi.org/10.1007/s00122-016-2788-x

Biehl, João and Peter Locke, eds. 2017. *Unfinished: The Anthropology of Becoming*. Durham, NC: Duke University Press. https://doi.org/10.1215/9780822372455

Bijker, Wiebe E. and John Law 1992. General Introduction. In *Shaping Technology/Building Society: Studies in Sociotechnical Change*. Wiebe E. Bijker and John Law, eds. Pp. 1–14. Cambridge, MA: The MIT Press.

Blackwell, Trevor and Jeremy Seabrook 1993. *The Revolt Against Change. Towards a Conserving Radicalism*. London: Vintage.

Blake, William 2011 [1793]. *The Marriage of Heaven and Hell*. Oxford: The Bodleian Library.

Bogue, Ronald 2016. *Deleuze's Way: Essays in Transverse Ethics and Aesthetics*. Aldershot: Ashgate Publishing. https://doi.org/10.4324/9781315576312

Bonneuil, Christophe and Frédéric Thomas 2009. *Gènes, pouvoirs et profits. Recherche publique et régimes de production des savoirs de Mendel aux OGM*. Paris: Quae.

Borup, Mads, Nik Brown, Kornelia Konrad and Harro van Lente 2006. The Sociology of Expectations in Science and Technology. *Technology Analysis and Strategic Management* 18(3/4):285–298. https://doi.org/10.1080/09537320600777002

Bourdieu, Pierre 1977. *Outline of a Theory of Practice*. Cambridge: Cambridge University Press. https://doi.org/10.1017/cbo9780511812507

Bovensiepen, Judith M. 2020. On the Banality of Wilful Blindness: Ignorance and Affect in Extractive Encounters. *Critique of Anthropology* 40(4):490–507. https://doi.org/10.1177/0308275X20959426

Braun, Kathryn. 2007. Biopolitics and Temporality in Arendt and Foucault. *Time and Society* 16(5):5–23. https://doi.org/10.1177%2F0961463X07074099

Bryant, Rebecca and Daniel M. Knight 2019. *The Anthropology of the Future*. Cambridge: Cambridge University Press.

Byerlee, Derek and Ken Fischer 2002. Accessing Modern Science: Policy and Institutional Options for Agricultural Biotechnology in Developing Countries. *World Development* 30(6):931–948. https://doi.org/10.1016/S0305-750X(02)00013-X

Candea, Matei and Lys Alcayna-Stevens 2012. Internal Others: Ethnographies of Naturalism. *The Cambridge Journal of Anthropology* 30(2):36–47. https://doi.org/10.3167/ca.2012.300203

Carmichael, Jim 2004. Apomixis for Crop Production: Status of Technology Development and Commercialization Implications. *Willamette Journal of International Law and Dispute Resolution* 12(1):29–48.

Carrère, Louis and Mario Pavese 2001. Screening Procedures to Identify and Quantify Apomixis. In *The Flowering of Apomixis*. Yves Savidan, John G. Carman, and Thomas Dresselhaus, eds. Pp. 121–136. Mexico DF: CIMMYT.

Carrère, Louis, Céline Dumesnil and Mariana Rivera 2002. Cell Cycle Progression during Endosperm Development in Zea mays Depends on Parental Dosage Effects. *The Plant Journal* 32:1057–1066.

Carrère, Louis, Roberto Valastro, Mariana Rivera, Pablo A. Galindo, Ana M. Soriano-Martinez and Félix Palomar 2009. Seed Development and Inheritance Studies in Apomictic Maize-*Tripsacum* Hybrids Reveal Barriers for the Transfer of Apomixis into Sexual Crops. *International Journal of Developmental Biology* 53(4):585–596.

Casañas, Francesc, João Simó, Joan Casals and Jaime Prohens 2017. Toward an Evolved Concept of Landrace. *Frontiers in Plant Science* 8:145. https://doi.org/10.3389/fpls.2017.00145

Casci, Tanita 2011. Biotechnology: Seeds of No Change. *Nature Reviews Genetics* 12:228–229. https://doi.org/10.1038/nrg2979

CGIAR 1996. Scientists Announce a Breakthrough in Research on 'Asexual' Maize. *CGIAR Newsletter* 3(2):1–3. Washington, DC: CGIAR.

Charles, Daniel 2003. Corn That Clones Itself. *Technology Review* 106:32–41.

Conner, Joann A. and Peggy Ozias-Akins 2017. Apomixis: Engineering the Ability to Harness Hybrid Vigor in Crop Plants. *Methods in Molecular Biology* 1669:17–34. https://doi.org/10.1007/978-1-4939-7286-9_2

Crouch, Jonathan 2009. Apomixis Consortium: Characterization of the Functional Components required for Apomixis in Maize. Mexico DF: CIMMYT Briefing Paper.

Curtis, Mark, Lukas Brand, Wei Yang, Eveline Nuesch, James M. Moore, Richard Jefferson and Ueli Grossniklaus 2004. Apomixis Technology Development: Transgene Containment and Fixation of Heterosis. In *Proceedings of the 8th International Symposium on the Biosafety of Genetically Modified Organisms*, Montpellier, France, 26–30 September: 167–173

De Beistegui, Miguel. 2010. *Immanence: Deleuze and Philosophy*. Edinburgh: Edinburgh University Press. https://doi.org/10.3366/edinburgh/9780748638307.001.0001

Deleuze, Gilles 1987. 'What Is Philosophy?' and the Notion of Substance. In *Leibniz and the Baroque. The Deleuze Seminars*. University of Paris Vincennes-St. Denis, 1986–1987. Lecture 15, 28 April 1987. Published online by Purdue University: https://deleuze.cla.purdue.edu/index.php/seminars/leibniz-and-baroque/lecture-15, accessed 28 August 2020.

Deleuze, Gilles 1988. *Bergsonism*. New York: Zone Books.

Deleuze, Gilles 1995. Control and Becoming. Conversation with Toni Negri. In *Negotiations, 1972–1990*. Pp. 169–176. New York: Columbia University Press.

Deleuze, Gilles 2004. *Difference and Repetition*. London: Continuum Books.

Deleuze, Gilles 2007. Contrôle et devenir. In *Pourparlers 1972–1990*. Pp. 229–239. Paris: Les Éditions de Minuit.

Deleuze, Gilles and Félix Guattari 1988. *A Thousand Plateaus: Capitalism & Schizophrenia*. London: Athlone Press.

Deleuze, Gilles and Félix Guattari 1994. *What Is Philosophy?* London: Verso Books.

Dressel, Dieter, Jim Carmichael and Thomas Éluard 2001. Genetic Engineering of Apomixis in Sexual Crops: A Critical Assessment of the Apomixis Technology. In *The Flowering of Apomixis*. Yves Savidan, John G. Carman, and Thomas Dresselhaus, eds. Pp. 229–243. Mexico DF: CIMMYT.

Dupré, John and Daniel J. Nicholson 2018. A Manifesto for a Processual Philosophy of Biology. In *Everything Flows: Towards a Processual Philosophy of Biology*. Daniel J. Nicholson and John Dupré, eds. Pp. 3–45. Oxford: Oxford University Press. https://doi.org/10.1093/oso/9780198779636.003.0001

Éluard, Thomas 1978. Le XIVème congrèes international de génétique et les recherches sur l'apomixie en 1978. Internal Report. Paris: ORSTOM.

Éluard, Thomas 1982. Embryological Analysis of Facultative Apomixis in *Panicum maximum* Jacq *Crop Science* 22:467–469.

Éluard, Thomas 1995. Les promesses de l'apomixie. ORSTOM Actualitiés 47. Paris: ORSTOM.

Éluard, Thomas 2000. Apomixis: Genetics and Breeding. In *Plant Breeding Reviews*. Jules Janick, ed. Pp.13–86. New York: John Wiley.

Éluard, Thomas 2001. Transfer of Apomixis through Wide Crosses. In *The Flowering of Apomixis*. Yves Savidan, John G. Carman and Thomas Dresselhaus, eds. Pp.153–167. Mexico DF: CIMMYT.

Éluard, Thomas 2007. Apomixis in Higher Plants. In *Apomixis: Evolution, Mechanisms and Perspectives*. E. Hörandl, Ueli Grossniklaus, Peter van Dijk and Timothy. F. Sharbel, eds. Pp. 15–22. Ruggell, Liechtenstein: ARG Gantner Verlag.

Éluard, Thomas 2020. L'OGM qui nous va bien. Self-Published.

Éluard, Thomas and Emmanuel Marceau 1994. Maize x *Tripsacum* Hybridization and the Potential for Apomixis Transfer for Maize Improvement. In *Biotechnology in Agriculture and Forestry*. Y. P. S. Bajaj, ed. Pp.69–83. Berlin: Springer-Verlag.

Éluard, Thomas, Roberto Valastro, Gabriel Atxaga and Louis Carrère 1997. Means for Detecting Nucleotide Sequences Involved in Apomixis. Patent Application WO 1998/036090 A1 (FR97/01821).

Engebrigtsen, Ada Ingrid 2017. Key Figure of Mobility: the Nomad. *Social Anthropology* 25(1): 42–54. https://doi.org/10.1111/1469-8676.12379

Fabian, Johannes 1983. *Time and the Other: How Anthropology Makes Its Object*. New York: Columbia University Press. https://doi.org/10.7312/fabi16926

Ferreira de Carvalho, Julie, Carla Oplaat, Nikolaos Pappas, Martijn Derks, Dick de Ridder and Koen J. F. Verhoeven 2016. Heritable Gene Expression Differences between Apomictic Clone Members in *Taraxacum officinale*:

Insights into Early Stages of Evolutionary Divergence in Asexual Plants. *BMC Genomics* 17:203. https://doi.org/10.1186/s12864-016-2524-6

Ferret, Carole 2012. Toward an Anthropology of Action. André-Georges Haudricourt and Technical Efficiency. *L'Homme* 202(2):113–139. https://doi.org/10.4000/lhomme.23041

Fortun, Mike 2008. *Promising Genomics: Iceland and deCODE Genetics in a World of Speculation.* Berkeley: University of California Press. https://doi.org/10.1525/california/9780520247505.001.0001

Foucault, Michel 1966. *Les mots et les choses.* Paris: Gallimard. https://doi.org/10.14375/np.9782070293353

Foucault, Michel 1977. *Discipline and Punish: The Birth of the Prison.* London: Allen Lane.

Foucault, Michel 1980. *Power/Knowledge. Selected Interviews and Other Writings, 1972–1977.* London: Harvester Wheatsheaf.

Foucault, Michel 1998. Life: Experience and Science. In *Aesthetics, Method, and Epistemology. Essential Works of Foucault 1954–1984, Volume Two.* James D. Faubion, ed. Pp.465–478. New York: The New Press.

Franklin, Sarah 2003. Re-thinking Nature-Culture: Anthropology and the New Genetics. *Anthropological Theory* 3(1):65–85. https://doi.org/10.1177%2F1463499603003001752

Franklin, Sarah 2007. *Dolly Mixtures: The Remaking of Genealogy.* Durham, NC: Duke University Press. https://doi.org/10.1215/9780822389651

Franklin, Sarah 2013. *Biological Relatives: IVF, Stem Cells, and the Future of Kinship.* Durham NC: Duke University Press. https://doi.org/10.26530/oapen_469257

Franklin, Sarah and Margaret Lock 2003. Animation and Cessation: The Remaking of Life and Death. In *Remaking Life & Death: Toward an Anthropology of the Biosciences.* Sarah Franklin and Margaret Lock, eds. Pp.3–23. Santa Fe: School of American Research Press.

Gault, Christine M., Karl A. Kremling and Edward S. Buckler. 2018. *Tripsacum De novo Transcriptome Assemblies Reveal Parallel Gene Evolution with Maize after Ancient Polyploidy. The Plant Genome* 11(3). https://doi.org/10.3835/plantgenome2018.02.0012

Gell, Alfred 1992. *The Anthropology of Time. Cultural Constructions of Temporal Maps and Images.* Oxford and Providence: Berg.

Giddens, Anthony 2002. *Runaway World: How Globalization is Reshaping Our Lives.* London: Routledge.

Gledhill, John. 1994. *Power and Its Disguises.* London: Pluto Press.

Glémin, Sylvain, Celine Scornavacca, Jacques Dainat, Concetta Burgarella, Véronique Viader, Morgane Ardisson, Sarah Gautier, Sylvain Santoni, Jacques David and Vincent Ranwez 2019. Pervasive Hybridizations in the History of Wheat Relatives. *Science Advances* 5(5):eeav9188. https://doi.org/10.1126/sciadv.aav9188

GRAIN 2001. Apomixis: the Plant-Breeder's Dream. *Seedling* 18(3).

Gurvitch, Georges 1964. *The Spectrum of Social Time.* Dordrecht: D. Reidel. https://doi.org/10.1007/978-94-010-3623-8

Hardt, Michael and Antonio Negri 2000. *Empire*. Cambridge, MA: Harvard University Press.

Harlan, Jack R., M. H. Brooks, D. S. Borgaonkar and J. M. J. de Wet. 1964. Nature and Inheritance of Apomixis in *Bothriochloa* and *Dichanthium*. *Botanical Gazette* 125(1):41–46. https://doi.org/10.1086/336242

Harwood, Richard R., F. Place, A. H. Kassam and H. M. Gregersen 2006. International Public Goods through Integrated Natural Resources Management Research in CGIAR Partnerships. *Experimental Agriculture* 42(4):375–397. https://doi.org/10.1017/S0014479706003802

Haudricourt, André 1962. Domestication des animaux, culture des plantes et traitement d'autrui. *L'Homme* 2(1):40–50. https://doi.org/10.3406/hom.1962.366448

Haudricourt, André 1964. Nature et culture dans la civilisation de l'igname: l'origine des clones et des clans. *L'Homme* 4(1):93–104. https://doi.org/10.3406/hom.1964.366613

Haudricourt, André and Louis Hédin 1943. *L'Homme et les plantes cultivées*. Paris: Gallimard.

Heidegger, Martin 1993a. *Basic Writings*. London: Routledge.

Heidegger, Martin 1993b. The Question Concerning Technology. In *Basic Writings*. Pp. 307–341. London: Routledge.

Heisey, Paul, Chittur S. Srinivasan, and Colin Thirtle. 2001. *Public Sector Plant Breeding in a Privatizing World*. Agricultural Information Bulletin No. 772. Washington, DC: United States Department of Agriculture, Economic Research Service.

Helmreich, Stefan 2007. Blue-Green Capital, Biotechnological Circulation and an Oceanic Imaginary: A Critique of Biopolitical Economy. *Biosocieties* 2:287–302. https://doi.org/10.1017/S1745855207005753

Helmreich, Stefan 2008. Species of Biocapital. *Science as Culture* 17(4):463–78. https://doi.org/10.1080/09505430802519256

High, Casey, Ann H. Kelly and Jonathan Mair, eds. 2012. *The Anthropology of Ignorance: An Ethnographic Approach*. New York: Palgrave Macmillan. https://doi.org/10.1057/9781137033123

Hodges, Matt 2008. Rethinking Time's Arrow: Bergson, Deleuze and the Anthropology of Time. *Anthropological Theory* 8(4):399–429. https://doi.org/10.1177%2F1463499608096646

Hodges, Matt 2019. History's Impasse: Radical Historiography, Leftist Elites, and the Anthropology of Historicism in Southern France. *Current Anthropology* 60(3):391–413. https://doi.org/10.1086/703204

Hojsgaard, Diogo 2020. Apomixis Technology: Separating the Wheat from the Chaff. *Genes* 11, 411. https://doi.org/10.3390/genes11040411

Husserl, Edmund 1966. *The Phenomenology of Internal Time-Consciousness*. Bloomington: Indiana University Press.

Ingold, Tim 2014. That's Enough about Ethnography. *Hau: Journal of Ethnographic Theory* 4(1):383–395. http://dx.doi.org/10.14318/hau4.1.021

IRD 1998. Décret n°84-430 du 5 juin 1984 portant organisation et fonctionnement de l'Institut de recherche pour le développement. Modifié 5 novembre 1998.

Jameson, Fredric 1998. *Brecht and Method*. London: Verso.

Jasanoff, Sheila 2005. *Designs on Nature: Science and Democracy in Europe and the United States*. Princeton: Princeton University Press. https://doi.org/10.1515/9781400837311

Jefferson, Richard 1994. Apomixis: A Social Revolution for Agriculture! *Biotechnology and Development Monitor* 19:14–16.

Khanday, Imtiyaz, Debra Skinner, Bing Yang, Raphael Mercier and Venkatesan Sundaresan 2019. A Male-Expressed Rice Embryogeneic Trigger Redirected for Asexual Propagation through Seeds. *Nature* 565:91–95. https://doi.org/10.1038/s41586-018-0785-8

Kindiger, Bryan, Viktor Sokolov and Chester Dewald 1996. A Comparison of Apomictic Reproduction in Eastern Gamagrass (*Tripsacum dactyloides*) and Maize-*Tripsacum* Hybrids. *Genetica* 97:103–110. https://doi.org/10.1007/BF00132586

Kingsbury, Noel 2009. *Hybrid: The History and Science of Plant Breeding*. Chicago: University of Chicago Press. https://doi.org/10.7208/chicago/9780226437057.001.0001

Kleinberg, Ethan, ed. 2012. Virtual Issue: The New Metaphysics of Time. *History and Theory*. https://onlinelibrary.wiley.com/page/journal/14682303/homepage/virtual_issue__the_new_metaphysics_of_time.htm, accessed 28 August 2020.

Kloppenburg, Jack 2004. *First the Seed: The Political Economy of Plant Biotechnology*. Madison: University of Wisconsin Press.

Koltunow, Anna and Ueli Grossniklaus 2003. Apomixis: A Developmental Perspective. *Annual Review of Plant Biology* 54:547–574. https://doi.org/10.1146/annurev.arplant.54.110901.160842

Koselleck, Reinhart 1985. *Futures Past: On the Semantics of Historical Time*. Cambridge, MA: MIT Press.

Lambek, Michael 2018. *Island in the Stream: An Ethnographic History of Mayotte*. Toronto: University of Toronto Press. https://doi.org/10.3138/9781487519049

Landecker, Hannah 2005. Living Differently in Time: Plasticity, Temporality and Cellular Biotechnologies. *Culture Machine* 7, an e-journal at: https://culturemachine.net/biopolitics/living-differently-in-time/, accessed 28 August 2020.

Landecker, Hannah 2007. *Culturing Life: How Cells Became Technologies*. Cambridge, MA: Harvard University Press. https://doi.org/10.4159/9780674039902

Latour, Bruno 1996. *Aramis, or the Love of Technology*. Cambridge, MA: Harvard University Press. https://doi.org/10.1628/978-3-16-156172-6

Latour, Bruno 2007. *Reassembling the Social: An Introduction to Actor-Network Theory*. Oxford: Oxford University Press.

Lebner, Ashley 2017. Introduction: Strathern's Redescription of Anthropology. In *Redescribing Relations: Strathernian Conversations on Ethnography, Knowledge and Politics*. Ashley Lebner, ed. Pp. 1–37. Oxford: Berghahn Books. https://doi.org/10.2307/j.ctvw04dqc.4

Legg, Stephen 2011. Assemblage/Apparatus: Using Deleuze and Foucault. *Area* 43(2):128–133. https://doi.org/10.1111/j.1475-4762.2011.01010.x

Léon-Martínez, Gloria and Jean-Philippe Vielle-Calzada 2019. Apomixis in Flowering Plants: Developmental and Evolutionary Considerations. *Current Topics in Developmental Biology* 131:565–604. https://doi.org/10.1016/bs.ctdb.2018.11.014

Lerman, Lindsay 2015. Georges Bataille's 'Nonknowledge' as Epistemic Expenditure: An Open Economy of Knowledge. PhD Thesis. Guelph, Ontario: University of Guelph.

Luhmann, Niklas 1993. *Risk: A Sociological Theory*. Berlin: Walter de Gruyter.

Lundy, Craig 2012. *History and Becoming. Deleuze's Philosophy of Creativity*. Edinburgh: Edinburgh University Press.

Lyman, R. Lee 2007. What Is the 'Process' in Cultural Process and in Processual Archaeology. *Anthropological Theory* 7(2):217–250. https://doi.org/10.1177%2F1463499607077299

MacKenzie, Donald and Judy Wajcman, eds. 1999. *The Social Shaping of Technology*. Buckingham: Open University Press.

McMeniman, S. and G. Lubulwa 1997. Project Development Assessment: an Economic Evaluation of the Potential Benefits of Integrating Apomixis into Hybrid Rice. Economic Evaluation Unit Working Paper No. 28. Canberra: ACIAR.

Marcus, George E. and Erkan Saka 2006. Assemblage. *Theory, Culture and Society* 23(2–3):101–109. https://doi.org/10.1177/0263276406062573

Marder, Michael. 2013. *Plant-Thinking: A Philosophy of Vegetal Life*. New York: Columbia University Press.

Marimuthu, Mohan P. A., Sylvie Jolivet, Maruthachalam Ravi, Lucie Pereira, Jayeshkumar N. Davda, Laurence Cromer, Lili Wang, Fabien Nogué, Simon W. L. Chan, Imran Siddiqi and Raphael Mercier 2011. Synthetic Clonal Reproduction through Seeds. *Science* 18:331(6019):876 https://doi.org/10.1126/science.1199682

Morson, Gary Saul 1994. *Narrative and Freedom: The Shadows of Time*. New Haven: Yale University Press.

Mounolou, J.-C. and A. Sarr 1990. Jean Pernès. *Medical Sciences* (Paris) 6(7):x.

Murphy, Denis 2007. *Plant Breeding and Biotechnology: Societal Context and the Future of Agriculture*. Cambridge: Cambridge University Press.

Myers, Natasha 2015. Conversations on Plant Sensing: Notes from the Field. *NatureCulture* 3:35–66.

Nas, Tevfik F. 1996. *Cost-Benefit Analysis: Theory and Application*. London: SAGE.

Nietzsche, Friedrich 1989. On Truth and Lying in an Extra-Moral Sense. In *Friedrich Nietzsche on Rhetoric and Language*. Sander L. Gilman, Carole Blair and David J. Parent, eds. Pp. 246–257. New York: Oxford University Press.

Noys, Benjamin 1998. Georges Bataille's Base Materialism. *Journal for Cultural Research* 2(4):499–517. https://doi.org/10.1080/14797589809359312

Nuffield Council on Bioethics 1999. Genetically Modified Crops: The Ethical and Social Issues.

O'Malley, Pat 2004. *Risk, Uncertainty and Government*. London: The GlassHouse Press. https://doi.org/10.4324/9781843146025

Ong, Aihwa and Stephen J. Collier 2005. *Global Assemblages: Technology, Politics, and Ethics as Anthropological Problems*. Oxford: Blackwell.

Osborne, Peter 1995. *The Politics of Time: Modernity and Avant-Garde*. London: Verso.

Ozias-Akins Peggy, E. L. Lubbers, Wayne W. Hanna and J. W. McNay 1993. Transmission of the Apomictic Mode of Reproduction in Pennisetum: Co-inheritance of the Trait and Molecular Markers. *Theoretical Applied Genetics* 85(5):632–638. https://doi.org/10.1007/BF00220923

Palmié, Stephan and Charles Stewart 2016. Introduction: For an Anthropology of History. Special Section, The Anthropology of History. *Hau: Journal of Ethnographic Theory* 6(1):207–236. https://doi.org/10.14318/hau6.1.014

Parr, Adrian, ed. 2010. *The Deleuze Dictionary*. New York: Columbia University Press.

Passerin d'Entrèves, Maurizio 1994. *The Political Philosophy of Hannah Arendt*. London: Routledge.

Paul, Helena and Ricarda Steinbrecher 2003. *Hungry Corporations: Transnational Biotech Companies Colonise the Food Chain*. London: Zed Books.

Pedersen, Morten Axel 2017. Infrastructure and Ignorance in Peri-Urban Ulaanbaatar. *Cambridge Journal of Anthropology* 35(2):79–95. https://doi.org/10.3167/cja.2017.350207

Penrose, David 1997. Apomixis, a Research Biotechnology for the Resource-Poor: Some Ethical and Equity Considerations. In *Intellectual Property Rights in Agriculture*. Uma Lele, William H. Lesser and Ges Horstkotte-Wesseler, eds. Pp.31–33. Washington, DC: World Bank.

Petrov, Dmitri Fedorovich 1984. *Apomixis and Its Role in Evolution and Breeding*. New Delhi: Oxonian Press.

Pickering, Andrew 1995. *The Mangle of Practice: Time, Agency and Science*. Chicago: University of Chicago Press. https://doi.org/10.7208/chicago/9780226668253.001.0001

Pickering, Andrew 2008. New Ontologies. In *The Mangle in Practice: Science, Society, and Becoming*. Andrew Pickering and Keith Guzik, eds. Pp.1–14. Durham, NC: Duke University Press. https://doi.org/10.1215/9780822390107

Pickering, Andrew 2010. *The Cybernetic Brain: Sketches of Another Future*. Chicago: University of Chicago Press. https://doi.org/10.7208/chicago/9780226667928.001.0001

Pickering, Andrew 2017. The Ontological Turn: Taking Different Worlds Seriously. *Social Analysis* 61(2):134–150. https://doi.org/10.3167/sa.2017.610209

Pollock, Andrew 2000. Looking for Crops That Clone Themselves. *The New York Times*, 25 April: F3.

Power, Michael 2007. *Organized Uncertainty: Designing a World of Risk Management*. Oxford: Oxford University Press.

Proctor, Robert N. and Londa Schiebinger, eds. 2008. *Agnotology: The Making and Unmaking of Ignorance.* Stanford: Stanford University Press.

Pynchon, Thomas 1973. *Gravity's Rainbow.* New York: Viking Press.

Rabinow, Paul 1996. *Making PCR: A Story of Biotechnology.* Chicago: Chicago University Press. https://doi.org/10.7208/chicago/9780226216874.001.0001

Rabinow, Paul 2003. *Anthropos Today: Reflections on Modern Equipment.* Princeton: Princeton University Press. https://doi.org/10.1515/97814008 25905

Rabinow, Paul 2008. *Marking Time. On the Anthropology of the Contemporary.* Princeton: Princeton University Press. https://doi.org/10.1515/97814008 27992

Rabinow, Paul 2011. *The Accompaniment: Assembling the Contemporary.* Chicago: University of Chicago Press. https://doi.org/10.7208/chicago/ 9780226701714.001.0001

Rainbird, Paul 2001. Deleuze, Turmeric and Palau: Rhizome Thinking and Rhizome Use in the Caroline Islands. *Journal de la Société des Océanistes* 112(1):14–19. https://doi.org/10.4000/jso.1653

Ransom, John S. 1997. *Foucault's Discipline: The Politics of Subjectivity.* Durham, NC: Duke University Press. https://doi.org/10.1215/9780822 382065

Rescher, Nicholas 2000. *Process Philosophy.* Pittsburgh: University of Pittsburgh Press. https://doi.org/10.2307/j.ctt6wrc3b

Rheinberger, Hans-Jörg 1997. *Towards a History of Epistemic Things: Synthesizing Proteins in the Test Tube.* Stanford: Stanford University Press.

Richards, Paul 2004. Private versus Public? Agenda-Setting in International Agro-Technologies. In *Agribusiness and Society: Corporate Response to Environmentalism, Market Opportunities and Public Regulation.* Kees Jansen and Sietze Vellema, eds. Pp.261–288. London: Zed Books.

Robbins, Joel 2001. Secrecy and the Sense of an Ending: Narrative, Time, and Everyday Millenarianism in Papua New Guinea and in Christian Fundamentalism. *Comparative Studies in Society and History* 43(3):525–551. https://doi.org/10.1017/S0010417501004212

Rose, Nikolas 2007. *The Politics of Life Itself: Biomedicine, Power, and Subjectivity in the Twenty-First Century.* Princeton: Princeton University Press. https://doi.org/10.2307/j.ctt7rqmf

Sailer, Christian, Bernhard Schmid and Ueli Grossniklaus 2016. Apomixis Allows the Transgenerational Fixation of Phenotypes in Hybrid Plants. *Current Biology* 26:331–337. http://dx.doi.org/10.1016/j.cub.2015.12.045

Schmidt, Anja 2020. Controlling Apomixis: Shared Features and Distinct Characteristics of Gene Regulation. *Genes* 11:329. https://doi.org/10.3390/ genes11030329

Schutz, Alfred 1967. *The Phenomenology of the Social World.* Evanston, IL: Northwestern University Press.

Schwartz, James 2008. *In Pursuit of the Gene: From Darwin to DNA.* Cambridge: Harvard University Press. https://doi.org/10.4159/9780674043336

Shreeve, James 2004. *The Genome War: How Craig Venter Tried to Capture the Code of Life and Save the World*. New York: Alfred A. Knopf.

Smith, John 1841. XXXII. Notice of a Plant Which Produces Perfect Seeds without any Apparent Action of Pollen. *Transactions of the Linnean Society of London* 18(4): 509–512. https://doi.org/10.1111/j.1095-8339.1838.tb00200.x

Smith, M. Estellie 1982. The Process of Sociocultural Continuity. *Current Anthropology* 23(2):127–142. https://doi.org/10.1086/202797

Spielman, David J. and Klaus von Grebmer 2004. Public-Private Partnerships in Agricultural Research: An Analysis of Challenges Facing Industry and the Consultative Group on International Research. EPTD Discussion Paper No.113. Washington DC: International Food Policy Research Institute.

Spielman, David J., Frank Hartwich, and Klaus von Grebmer 2006. Public-Private Partnerships in International Agricultural Research. Washington, DC: International Food Policy Research Institute Research Brief No. 9.

Spillane, Charles, Jean-Philippe Vielle-Calzada and Ueli Grossniklaus 2001. APO2001: A Sexy Apomixer in Como. *The Plant Cell* 13(7):1480–1491. https://doi.org/10.1105/tpc.13.7.1480

Spillane, Charles, Mark D. Curtis and Ueli Grossniklaus 2004. Apomixis Technology Development—Virgin Births in Farmer's Fields? *Nature Biotechology* 22(6):687–691. https://doi.org/10.1038/nbt976

Stasch, Rupert 2009. *Society of Others: Kinship and Mourning in a West Papuan Place*. Berkeley: University of California Press. https://doi.org/10.1525/9780520943322

Stitzer, Michelle C. and Jeffrey Ross-Ibarra 2018. Maize Domestication and Gene Interaction. *New Phytologist* 220(2):395–408. https://doi.org/10.1111/nph.15350

Stone, Glenn Davis 2014. Biosecurity in the Age of Genetic Engineering. In *Bioinsecurity and Vulnerability*. Nancy N. Chen and Lesley A. Sharp, eds. Santa Fe: School for Advanced Research Press. Pp.71–86.

Strathern, Marilyn 1987. Out of Context: The Persuasive Fictions of Anthropology. *Current Anthropology* 28(3):251–281. https://doi.org/10.1086/203527

Strathern, Marilyn 1988. *The Gender of the Gift. Problems with Women and Problems with Society in Melanesia*. Berkeley: University of California Press. https://doi.org/10.1525/9780520910713

Strathern, Marilyn 2017. Gatherer Fields. A Tale about Rhizomes. *Anuac: Rivista della Società di Antropologia Culturale* 6(2):23–44. https://doi.org/10.7340/anuac2239-625X-3058

Sunder Rajan, Kaushik 2006. *Biocapital: The Constitution of Postgenomic Life*. Durham, NC: Duke University Press. https://doi.org/10.1215/9780822388005

Talcott, Samuel 2019. *Georges Canguilhem and the Problem of Error*. New York: Palgrave Macmillan. https://doi.org/10.1007/978-3-030-00779-9

Valastro, Roberto, Joe Tohme and Diego González-León 2001. Applications of Molecular Genetics in Apomixis Research. In *The Flowering* of Apomixis.

Yves Savidan, John G. Carman and Thomas Dresselhaus, eds. Pp.83–94. Mexico DF: CIMMYT.

Van Dijk, Peter 2008. Biotechnology: A Hold on Plant Meiosis. *Nature* 451, 28 February 2008: 1063–1064. https://doi.org/10.1038/4511063a

Van Dijk, Peter and Jos M. M. van Damme 2000. Apomixis Technology and the Paradox of Sex. *Trends in Plant Science* 5:81–84. https://doi.org/10.1016/S1360-1385(99)01545-9

Van Dijk, Peter, Diana Rigola and Stephen E. Schauer 2016. Plant Breeding: Surprisingly, Less Sex is Better. *Current Biology* 26:R102–R124. http://dx.doi.org/10.1016/j.cub.2015.12.010

Van Dijk, Peter, Hans de Jong, Kitty Vijverberg and Arjen Biere 2009. An Apomixis-Gene's View on Dandelions. In Isa Schön, Koen Martens and Peter van Dijk, eds. *Lost Sex: The Evolutionary Biology of Parthenogenesis.* Dordrecht: Springer. https://doi.org/10.1007/978-90-481-2770-2

Viveiros de Castro, Eduardo 2014. *Cannibal Metaphysics. For a Post-Structural Anthropology.* Minneapolis: University of Minnesota Press.

Viveiros de Castro, Eduardo 2015. *The Relative Native: Essays on Indigenous Conceptual Worlds.* Chicago: Hau Books.

Whitehead, Alfred North 1979. *Process and Reality.* New York: Free Press.

Williams, Robin and David Edge 1996. The Social Shaping of Technology. *Research Policy* 25: 865–899. https://doi.org/10.1016/0048-7333(96)00885-2

Winkler, Hans 1906. Botanische Untersuchungen aus Buitenzorg. II, 7. Ueber Parthenogenesis bei *Wikstroemia indica* (L.). C. A. Meyer. *Annales du Jardin Botanique de Bruitenzorg* 20(2,5):208–276.

Wolf, Eric 1982. *Europe and the People Without History.* Berkeley: University of California Press.

Archival sources

AR 1993: CIMMYT Annual Report, 1993

AR 1994: CIMMYT Annual Report, 1994

ER 1994: External Review of the ORSTOM-CIMMYT Apomixis Project, CIMMYT 1994.

IR 1994: Internal Review of the ORSTOM-CIMMYT Apomixis Project, CIMMYT, 1994

IR 1997: Internal Review of the ORSTOM-CIMMYT Apomixis Project, CIMMYT, 1997

PW 2002: CIMMYT Project Week 2002, Part 4: Frontier Project Reports. CIMMYT 2002.

Index